The View From Here

Optimize Your Engineering Career From the Start

Reece Lumsden

Foreword by Dr. Patricia Galloway

The View From Here
Optimize your engineering career from the start

Copyright ©2018 Reece Lumsden
All rights reserved.

Printed in the United States of America.

No portion of this work may be reproduced or transmitted in any form or by any means, electronic or mechanical, including photocopying, scanning, and recording, or by an information storage or retrieval system without written permission from Reece Lumsden.

SECOND EDITION

Cover & Interior design by W. Bruce Conway

Cover photo courtesy of MS Clipart Collection

Printed on recycled paper.

Soft Cover ISBN 197-6-419-786

Library of Congress Control Number: 2010942771

Praise For the *The View From Here by Reece Lumsden*

Every generation needs mentors, people who are willing to invest in the healthy growing up of a new generation. The pace of society has increased so much over the past decades that the youth of today are often left unmentored because of lack of time. Reece's writing and insight is bridging this gap by answering some of the pertinent questions facing engineering students and young engineers today. I encourage budding engineers to learn from Reece and in turn impart your experience to the next generation.

Dr. D. Chong—Vice-Chair, Student Activities Committee
Institute of Electrical and Electronics Engineers (IEEE)

Engineering Education has emerged as an important academic field that helps in transforming eager students into capable engineers. Despite the rise of Engineering Education as a recognized discipline, a generation gap seems to divide the young person with a technical bent from the mature engineer in industry. Throughout an engineering student's academic educational process, uncertainties about the new engineer's professional transition hinder the mental clarity and personal commitment of the uninitiated. The difficulties are reflected in the fact that enrollment in U.S. engineering schools is not growing as would be expected in the environment of a growing population, increased wage pressures, and exponential increase in the use of technology.

The View From Here excellently addresses the engineering student's passage through professional barriers that remain opaque despite decades of maturation in engineering practice, guiding the novice through the labyrinths of previously only experience-based passages that transform the lay person into a practicing engineer. Reece Lumsden's conversational tone facilitates not only the absorption of knowledge but, more importantly, a rise in confidence in the developing engineer. *The View From Here* will undoubtedly find an important place in engineering academic advising, as well as gain a reputation as an accessible and popular guide among engineering student social networks.

Dr. E. Smith—Systems Engineering Program
University of Texas-El Paso

As the 2008 – 2009 Power & Energy Society President, I have discussed the engineering profession with numerous students and engineers from around the world. *The View From Here* is a terrific resource for those contemplating an engineering career, getting their degree, looking for employment or already in the workforce. It provides advice for success throughout the engineering journey and offers exercises to help map your own destiny. It is practical, insightful and written with global perspective. I highly recommend it.

Ms. W. Reder—Vice President
Power Systems Services
S&C Electric Company

This book is a must-read for all graduating engineers. Engineers focus on understanding technical concepts, not necessarily understanding the corporate landscape. This book provides sound advice and excellent guidance to help young engineers navigate through their careers.

Dr. D. Cassone—Logistics Manager
Sprint Nextel

This is not a book just for younger engineers or even engineers. It is great read for lay people and especially Baby Boomer executives who need to understand the newer generations they now manage. Reece addresses the dynamic tension between 20th century management styles banging into 21st century generational needs, the realities that result and how to cope. He gives valuable insight on the role of mentoring, bosses, the changing role of life versus work, balancing life's pressures, attitudes and more.

Mr. P. Gartz—Past President
Aerospace and Electronic Systems Society

I would highly encourage all engineering students to read and absorb the suggestions offered and use them as guidance to learn from the mistakes that many of us "older" engineers have made.

Dr. P. Galloway—CEO
Pegasus-Global Holdings
Author of The 21st Century Engineer

There are many work-related lessons that university doesn't teach you. Important lessons that every engineer needs to learn early on, as they determine who will succeed and advance in their careers and who will not. This book will provide you with advice to help you chart your own path. I recommend this excellent guide book to my students and junior engineers!

Dr. D. Musielak—Research Professor of Aerospace Engineering

Are you headed in the right direction? This book captures those first few critical years of an engineering career from the unique perspective of a young engineer. The chapter summaries, extensive practical examples and 'Actions to Take Away' make it instantly accessible and directly applicable. *The View From Here* is essential reading for every engineering student or young graduate. Read it and get ahead of the pack in the race for a great engineering career.

Mr. G. Walters—Tasmanian State President
Engineers Australia

Dedications

I dedicate this book to my wife Nik, and my parents Carol and Charles, all of whom have contributed in their own way and have been through a lot watching me try to get this book completed after 6 long years. This is as much your achievement as it is mine...

Reece Lumsden
December, 2010

Table of Contents

Foreword 9

Introduction 11

1. What is engineering? 13
2. Engineering and university 25
3. The Working World 41
4. So just what are employers looking for? 59
5. Job Hunting 71
6. Being an engineer—so what's it like? 89
7. Communication—the engineer's lynchpin 103
8. Outsourcing and off-shoring 119
9. So why aren't engineers… 153
10. The successful engineer's secret—attitude 161

Afterword 177

Acknowledgements 179

References 181

Further Reading 187

Bibliography 191

About the Author 193

Foreword

Engineering may truly be the world's oldest profession. The human race simply could not survive without engineers. Everything around us involves engineering. Reece Lumsden, in his new book, *The View From Here* clearly defines engineering and why it is so important. He raises the critical issue of reduced university engineering enrollment and why the time is now to encourage and excite young people to consider engineering as a career. Lumsden offers students considering and studying engineering an opportunity to get an early start on their careers by laying out the needed skill sets to become "professionals", and not "technicians" the public sometimes perceive we are. He provides insights as to what employers look for when interviewing engineering candidates and provides a guide on how to demonstrate that he or she possesses the right skill sets.

Lumsden offers approaches for helping engineering students prepare for the future in the face of uncertainty, including seeking mentors who have "been there-done that." Lumsden goes further, noting that mentoring is a two-way street and that employers must also reach out to their young engineers to guide them; while at the same time providing an environment that best meets their younger generation life styles. His step-by-step guide to help one identify what skill sets may be required, what potential engineering careers may be of interest, and how to seek out companies or locations is a brilliant tool that I am sure many of us "older" engineers wish we had available to us when we were in engineering school.

We all recognize that our universities need to continue to teach engineering fundamentals, but if we are to produce engineering "professionals" that possess the skill sets employers seek, we must do much more. The bottom line is that engineers must be well-rounded and have additional skills if they are to justify the higher wages and respect they hope to command. Lumsden has correctly identified the skill sets required of 21st Century engineers including team work, communication, leadership, problem solving, systems thinking, business appreciation, time management, and *passion*. In discussing these skills sets, he lays out a road map of how best to proceed in seeking that perfect job, while aligning with the questions posed in the beginning of his book. In doing so, he offers thoughts for consideration as to what it means to be an engineer providing his own experiences of what it is like to be one and identifying advantages and disadvantages of various options one may choose when considering an engineering career.

I applaud Lumsden for identifying the most important skill to possess and yet is the skill that the public and employers alike say engineers don't possess-communication. His description of communication being the engineer's "lynchpin" is a statement to which I completely agree. I have lectured and written on the importance of communication for over three decades and believe it should be the top priority of engineers. Being the fundamental skill required from all roles and at all levels of an organization, Lumsden presents a concise summary of why communication is important and discusses various forms of communication including the pitfalls one may encounter if he or she does not recognize its importance.

Lumsden and I further agree that engineering is not well understood by the public and is not viewed the same as the medical and legal professions. As engineers, we have a long way to go in changing our image. In his book, he offers some interesting ideas of how young engineers today might change that image. It's all about the attitude we have. As Lumsden so eloquently states, *"Your attitude will impact your success as an engineer just as much as the tangible skills you bring to the table."*

His book comes at a key moment, challenging engineering students, universities and employers with ideas and suggested steps that can be taken to move the engineering profession to a higher plateau. As Lumsden reminds us, at some point in our professional career, most of us inevitably stop and wonder where it's all going and whether we are truly headed in the right direction. His book offers some of the best guidance I have seen on how best to answer those questions. Reece Lumsden is to be congratulated for writing a much-needed book for engineering students today. I would highly encourage all engineering students to read and absorb the suggestions offered and use the suggestions as guidance to learn from the mistakes that many of us "older" engineers have made. In doing so, you will have a better awareness of what issues an engineer faces in the world marketplace today and can prepare yourself for truly becoming a professional that can make a difference.

Dr. Patricia D. Galloway, PE, CPEng, PMP, MRICS

Chief Executive Officer, Pegasus-Global Holdings

Past President, The American Society of Civil Engineers (ASCE), 2004

Author of *The 21st Century Engineer*

Introduction

The idea for this book has been with me for a while now, but I've always been a little unsure how to go about it. Like most young engineers who care about their career, I've heard countless talks, lectures and seminars on what the engineering playing field was like. There was however, always one problem; those who were delivering the message weren't in my shoes. They couldn't tell me what it looked like from the perspective of an early career engineer; those usually in their twenties and early thirties. So I decided to do something about it.

The idea for this book solidified in late 2003 when I attended the Australian Engineering Excellence Awards. That night at the awards dinner, I talked with a fellow National Young Engineers Committee (NYEC) member at length on a range of engineering issues. By the end of the night, I realized I could no longer sit on my hands; my thoughts had to be put on paper, and if I needed any further prompting, the e-mail I received the following day (shown below) solidified my resolve.

To: Reece

Subject: career

Reece,

I really wanted to let you know how much I appreciated the discussion we had on Friday night at the awards. Hmm where to start, I do feel a tad silly even writing this, I'll keep though!

Umm, the conversation was regarding me not really knowing what to do etc, I have had the conversation with friends that I went to uni with, and we always just end up chasing our tails, and deciding that its too hard. This time was different, perhaps because I was forced to explain myself to someone not from my industry, and to someone who may not know me all that well.

Anyway, my main point is thank you very much, you could have quite easily just left the questioning at She'll be right mate, and you didn't. For me it was one of those experiences, where you have been considering something for so long, and then all of a sudden can see what it really means...

Anyway, thank you, I doubt you remember, but for me, I needed to let you know that you are in fact an amazing and passionate person, (people passionate and work passionate), and I know I am really lucky to have met you.

Thanks, Jen

At some point in their professional career, most people inevitably stop and wonder where it's all going; they ask themselves, "Am I headed in the right direction or is it all spiralling out of control? What warning signs or indicators are there to guide me?" It's akin to being inside a barrel in calm waters just before going over the edge of a waterfall --- you're not aware of the impending doom ahead because, based on what you're experiencing in the present, you have no way to know what's up ahead.

Books written by those who have travelled similar paths are a common means by which others gain some sense of what's ahead. That is my intent for this book; to tell you how the first few years of my engineering career have unfolded, and what I have learned from them --- to paint a picture of what my path has shown me.

There are not enough books currently available that speak to the true concerns of young engineers. No one person can have all of the answers, and I certainly don't pretend to. Some of the subject matter you will agree with, some you will not. I have tried where possible to substantiate my claims with facts and data, but in some cases the issues I raise can only be discussed through the lens of my own experience. Subsequently, this book is not meant to be a textbook or prescriptive in any way, rather it is intended to give you an awareness of the issues you may face as an engineer.

My intention is to connect with you, the reader, as though we are two engineering colleagues having a conversation. I hope you find this material useful and I'd appreciate any feedback you may have on it (good or bad) or any other issues that are a part of this subject matter but not explicitly addressed herein.

Reece Lumsden
reece@theviewfromherebook.net
www.theviewfromherebook.net

1. What Is Engineering?

> This chapter will focus on what engineering is, some of the high level traits of an engineer, and society's role in shaping these. Engineering and science are typically lumped together in the media, but engineers and scientists are different. While most people would agree that engineering is important, exactly why that is may not be entirely clear. One way to understand this importance is by looking at how many engineers there are, which we will do in this chapter.

Defining our terms

As any good engineer would do upon entering a professional discussion, let's define the terms we'll be using. It's important to do this so that we are aware of the common terms of reference.

The term "engineering" is defined as:

"The art or science of making practical application of the knowledge of pure sciences, such as physics, chemistry, biology, etc." [1]

Extending this, we define one who practices engineering, "an engineer," as:

"One versed in the design, construction and use of engines or machines, or in any of the various branches of engineering: a mechanical engineer, electrical engineer, civil engineer, etc." [2]

On the first day of first year engineering, they should put this definition up on the blackboard (or whiteboard) so if you think you're studying law or history, you'll know you're in the wrong class. For many of you, this may have been the first time you've actually seen a definition of what engineering is, and what is meant by the term "engineer." Not once throughout my engineering studies did I receive a straightforward explanation of what engineering, or an engineer, was.

If you were to tell someone you were an "engineer," it would be logical for them to assume you practice engineering. After all, a lawyer practices law and an accountant does accounting. But if we try to extend the same logic, we come to a bit of a problem. In comparing the definition of an engineer to what one actually does as an engineer, we find there's an incongruity.

You may think this is just semantics, and that I'm only playing with words, but as I mentioned in the preface, this book intends to explore ideas and issues not normally talked about, and the definition of an engineer is just such an issue. While there is nothing incorrect in the above definition of engineer, it is fairly limited. When I read that definition I envision a mechanic or electrician up to their elbows in grease or fiddling with wires. This description may have been sufficient in the 1950s and 1960s when engineers held a close affinity with technicians (such as electricians and mechanics), but in today's world, engineering encapsulates a far greater breadth.

With the onset of computers and technology, engineering has taken on a persona much broader than that portrayed in our dictionary definition. In today's world, it is a far more information-rich endeavor involving much more "paper" engineering than ever before. Although one can still build scale models and physically test designs, the increased processing power and sophistication of modern computers enables us to simulate and test increasingly complex phenomena. This can benefit us by reducing the kind of costly experimentation that comes with testing out hypotheses in tangible form. I realize all too well that a simulation is just that, a representation of a real world situation, and that sometimes there is no substitute for seeing how things behave in the real world. It would be unreasonably expensive, however, if we had to build a separate bridge for each mode of failure we desired to test.

This massive increase in the use of technology in our jobs has led to the term "knowledge worker." As Margot Caines, a leadership strategist, commented: "We are connected 24/7. Ninety percent of us are knowledge workers."[3] This alludes to the fact that with the addition of engineers who work on the shop floor, greater numbers of us are becoming office workers whose main domain is the computer, but whose consideration and scope take in a much broader purview than it did before.

This notion of the knowledge worker whose primary output is intellectual capital is not typically the way we are portrayed, or may think of ourselves, as engineers. The typical view is that our work is defined by the tangible outputs we produce, like the bridge or plane we build. But think about it for a moment: the engineer doesn't actually build the bridge or plane --- the technicians do. The engineer may design it, but the transformation from intangible design to tangible form is the job of the contractor and his or her team of technicians (welders, laborers, electricians, etc.). You can start to see that what the engineer has actually created is intellectual capital.

This hypothesis is substantiated by the historical derivation of the word "engineering." "Engineer" comes from the Latin "ingenium," which literally translated means "talent, genius, cleverness, or native ability."[4] Without these, all an engineer can do is follow a linear mode of thought, action and reaction --- not very desirable in an increasingly complex and non-linear world. Therefore, this is the reason why an engineer's greatest assets are the creativity and innovation he or she can bring to bear on any given task.

Why is engineering important?

Why is engineering so important to us? We need only look at the definition of engineering introduced earlier: the application of knowledge from the other pure sciences. This tells us engineering is purpose driven; it has a focus or outcome. You don't study physics just for the sake of it, but so you can apply it to solve some kind of real world issue or problem. That is precisely the distinction that separates engineers from scientists.

Since physics governs the world we live in, engineering can be applied to a range of situations that occur in our everyday (and not so everyday) lives. Everything we see around us in the built environment, from the cars we drive to the water we drink to the food we eat, involves engineering input throughout the product's lifecycle.

You might think that if engineering is so important, then it is probably quite old, and indeed, looking back through time we can see the results of its application.

The Pyramids

While still thought by some to have been created by alien ancestors, the pyramids were in fact the product of knowledge and application of engineering principles. The Egyptians (or aliens, to be respectful of those who subscribe to this view) may not have considered the pyramid's complete lifecycle, including in-service support, disposal, and environmental footprint. Nevertheless, the fact that they are still standing today, some 4000 years after their construction, stands as a testament to the impact rigorous application of engineering principles can have on the tangible realization of an idea.

This is but one of a myriad of examples extending back through time that demonstrate the importance of engineering and how it has created the world we live in.

So back to our original question: why is engineering so important? Let me offer this: there is not one element of our lives today unaffected by engineering; humans simply could not have evolved (industrially, not biologically) to the point we are at without its constant and continual use.

So just how many engineers are there?

Given engineering's prevalence in our everyday lives, it would be reasonable to assume one could easily catalogue the engineering population. Unfortunately, it's not. For one thing, engineers are so pervasive that it's very difficult to know where to start. It's also very difficult to define just who's doing engineering work and who isn't. How do you classify it? Is it design, operational support, continued maintenance, or disposal? Is it all of these? And what of other positions, such as Project and Engineering Management, where the perception of whether you're doing real engineering work or not depends on who you talk to? It's very hard to get an accurate figure.

Even obtaining engineering graduation levels from universities is not as easy as you might assume. For example, it can be difficult to compare the graduating figures of one country with that of another because the term "engineer" is interpreted differently, thereby providing a basis for distortion in the numbers.

How many engineers?

A Business Week online article by Vivek Wadhwa discovered that engineering figures reported for China lacked an element of consistency. Those graduating as motor mechanics and technicians, as well as those with two and three year associates degrees, were being included in figures for graduating engineers.[5]

Referring to the table of Engineers graduated for 2003 (next page), it should be realized that actual figures for the top three --- China, U.S. and India --- may be reduced if we just consider those graduating with four year engineering degrees, as is the case for the rest of the list. Based on this criteria, the U.S. graduates around 60,000 to 70,000 engineers per year.

As these figures yield only quantitative data, they cannot tell us the quality of those graduated or how they are used. Any argument

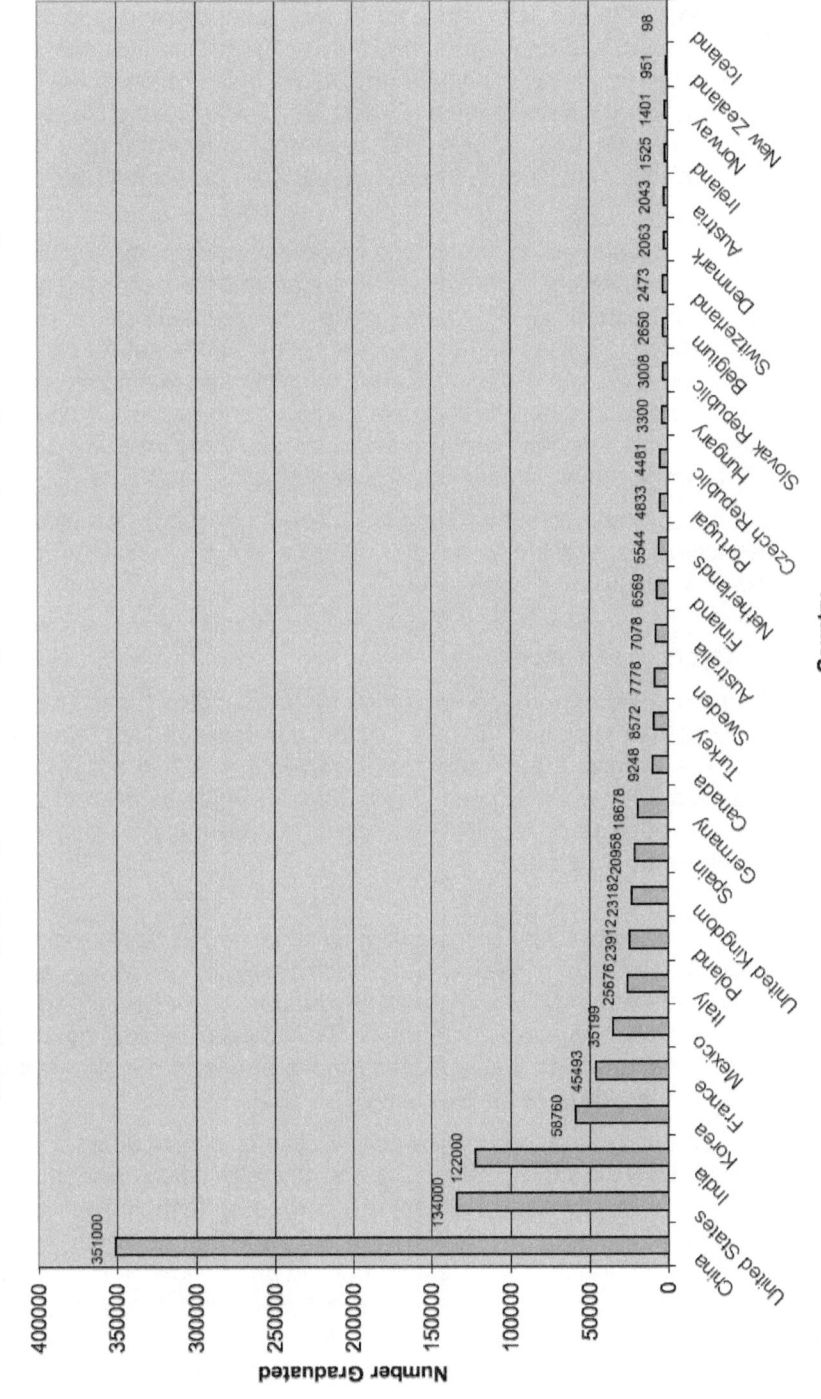

about engineering graduation based on pure figures alone is therefore nonsensical. In support of this, for China, despite graduating the largest number of engineers annually, it is quality, not quantity, that will determine the country's continued growth. As outlined in the 4th quarter 2005 edition of the McKinsey Quarterly, "While university graduates are plentiful [in China], new MGI research shows that only a small proportion of them have the skills required for jobs further up the value chain."[6]

For developed countries, the basic components of subsistence, such as the provision of food, clean water and the necessities of life, have been sorted out. Therefore, while engineering is still used to extract greater efficiencies in these areas, an additional focus can be placed on solving more complex issues, such as creating bionic ears, developing technology to map the human genome, and creating advanced materials that can withstand atmospheric re-entry. In other words, engineering is used to solve higher order problems.

Developing countries pursuing these higher order problems can be viewed as seeking to develop their country's prosperity from the top-down, instead of bottom-up.

Algeria and space

Algeria elected to become part of the five satellite Disaster Monitoring Constellation (DMC)[7]. While not regarded as having a space program, they saw that involvement in this program afforded them a way to gain useful benefits --- disaster monitoring being but one of many --- that could not be gained using other bottom-up methods.

Contrast this with the situation experienced in developing countries. These nations seek to use engineering to solve fundamental infrastructure needs largely taken for granted in the developed world, such as clean water, arable land and the provision of housing. Without solving these issues first, it doesn't make sense to develop capabilities to solve higher order problems.

Such a separation between developed and developing countries hints at the extent to which having a technically literate workforce can directly impact a nation's prosperity. But being technically literate is only half of the equation. As I discussed earlier, engineering is about creativity and innovation. It is about taking an application for a product or process used in one area somewhere else previously not thought of. It's about being observant, inquisitive, and constantly asking how something can be done faster, better, more efficiently and/or cheaper.

So how do I know if I'm cut out to be an engineer?

You may surmise that, by talking about such traits, I am alluding to common traits shared by all engineers. If you think you may like to do engineering, then you will probably fit a certain profile. I say "probably" because there are, of course, exceptions to the rule. However, those people who go into engineering will usually share the following attributes:

- Have a like for science and math
- Enjoy problem solving of one form or another
- Have an innate curiosity and enjoy finding out how things function

Concerning the first item, you can have a like for science and math yet not be good at it. How can that be? Surely, you might say, if you like something it's because you're good at it. People who like running or soccer are those who are good at it, otherwise, what's the point in doing it?

I'll illustrate what I mean with an example from my past. All throughout high school and university I enjoyed mathematics. However, I was not (and still am not) a natural at it. I was average, never receiving consistently high marks in the subject. I enjoyed it primarily because it involved solving problems which, to me, was a challenge. While I was happy to get correct answers, I far more enjoyed looking "underneath the hood" of a problem and figuring out what it was really all about. I saw the solving of problems as the real reward, not just a means to getting a right answer.

When students turn off to science or engineering early on, it's usually because of an aversion to difficulty. They haven't looked beyond solving a problem to the reward it will give them; they simply see it as a means to getting a right answer. They become frustrated and feel like a failure if they can't solve a problem. At this stage, for whatever reason, they are more concerned about the answer than why something is the way it is. They think if they don't get straight A's, they aren't any good at the subject. This is a false deduction: those not naturally talented at math and science often appreciate their successes more than others. They learn better from their mistakes and may be more disciplined when trying to master a craft for which they don't have a natural aptitude.

Further support for these views can be found in the studies conducted by Patricia Smiley and Carol Dweck, chronicled in Guy Claxton's book, *Wise Up—The Challenge of Lifelong Learning*. They investigated what it was that undermined resilience in young children. Essentially, they found that children who were more concerned about looking good to their parents or other authority figures were not terribly resilient. They either chose tasks so easy they knew they could complete them, or they removed any possibility of failure by shying away from difficult tasks. Claxton summed up this aversion to difficulty and appearance of success at all costs: "to be right is to be bright, and to be bright is to be good."[8]

Beginning early—how society de-sciences children

In our attempt to create more scientists and engineers, we must be aware that conditioning begins very early. We all start off as explorers. The world is one big mystery for a child, and every waking moment is spent finding out how everything works. Have you ever watched a small child? When they pick up a new toy for the first time they'll use all of their senses to figure out what it is, its purpose, how it sounds, what it feels like and of course, if they can put some part of it in their mouth, it'll get the taste test. They continue to do this, not vetting their behavior until they are able to walk and talk --- then they start running into problems.

Gradually, imperceptibly, most children arrive at a point where society's expectations have so reduced their view of themselves as explorers that they lose that inquisitive twinkle in their eye. When formal education comes around, children discover that asking questions is not always met with a positive response. Questions like "does it hurt bread when you bite it?" [9] or "why is the sky blue?" are not only part of the learning process for a child, but they also demand clarity from those answering the question as it tests their own understanding.

There's an expression that something is "so simple even a child can understand it." This is somewhat of a paradox, because while a child's understanding of the world is very simple, this very simplicity forces those answering questions to check their bravado at the door. A child continually asks "But why" until an adult provides an answer that makes sense. In most instances, the adult will try to cover their own lack of understanding by saying "Because I said so," or "That's enough."

Unfortunately, learning such patterns early on can establish an ingrained behavior that extends through adulthood, limiting the recipient to a learning environment that rarely extends beyond their intellectual shoreline because of the aversion they have learned to associate with asking questions. Clearly, if we are to have more children mature into people who retain a healthy curiosity and scepticism about the world around them, we need to let them know there are no silly questions.

The best scientific and engineering minds share a persistent level of inquisition and thirst for knowledge. New developments and breakthroughs in science and engineering will only come about from perseverance and the precedence of knowledge and understanding over being right and looking good.

Having established that widespread technical literacy is beneficial, we now need to consider how that literacy is impacted by the educational process. We shall therefore look more closely at the university environment and how it prepares us as engineers.

Chapter 1, if you only have 5 minutes...

- Engineering is the art or science of making practical application of the knowledge of pure sciences, such as physics, chemistry, biology.
- An engineer is one versed in the design, construction and use of engines or machines, or in any of the various branches of engineering: mechanical, electrical engineer, civil, etc.
- The definition of an engineer provides a thin view of the activities that an engineer is involved in.
- Engineering includes the ability to test increasingly complex phenomena through simulation, which saves time, money, and facilitates analysis.
- The main output of the engineer is intellectual capital.
- An engineer's greatest asset is their creativity and innovation.
- Engineering is different from science in that it is used to solve problems; it is purpose-driven.
- The use of engineering principles extends back to the earliest human endeavors, the Pyramids being one such example.
- Engineering permeates every facet of modern life, from the cars we drive to the food we eat.
- It is difficult to ascertain exactly how many engineers there are, given they work in a multitude of capacities across numerous fields.
- It is difficult to compare raw figures of engineering graduation numbers between countries, as there is no international standard definition of what an engineer is.
- Most engineers share common traits: they like science and math, enjoy problem solving of one form or another, possess an innate curiosity, and enjoy finding out how things function.
- Children begin life as explorers, one of the hallmarks of what makes a good engineer.

Actions to take away

Some things you can do to act on the information presented here:

Getting to understand engineering:

1. **Construct** your own definition of engineering based upon what you have experienced/read/learned so far.

2. **Search** for a definition of engineering from at least three different sources. Are there any differences among the definitions and with the definition you have developed? Why do you think this is?

3. **Comment** on the statement: "An engineer's greatest asset is their creativity and innovation." Do you agree? Why/Why not?

4. **Ask** yourself the following questions:

    a. Do I like science and math? Why/Why not?

    b. Do I enjoy problem solving? Why/Why not?

    c. Am I innately curious and do I enjoy finding out how things function? Why/Why not?

    d. Based on these answers, what can you say about your suitability for engineering?

The View From Here

2. Engineering And University

> In this section, we'll focus on some of the issues that surround studying engineering at university. The media, parents and university prestige all serve to exert influence on the decision of which university to study at. Once a school is decided on and you've gained acceptance, there are still further issues to grapple with. First you need to consider where a lecturer's focus lies: is it on teaching you, publishing papers, or something else? Then you have to deal with grading policies, cheating, and the contentious issue of how the university experience is viewed, whether it be a business or social contract. The utility of higher degrees, in particular the MBA, also gains more and more prevalence as engineers look to distinguish themselves from the crowd. We'll touch on the university's role (if any) in preparing engineers, and whether it should be considered an institution of pure learning or one focused on providing workplace ready engineers. Finally, we'll cover professional societies and how they fit into the university experience.

Running the gauntlet

When deciding to study engineering, just settling on which school to attend can be a major undertaking in itself. This decision will be influenced by many sources; the media, friends and family, alumni, career advisors and, of course, your grades. As the competition to gain acceptance to sought-after institutions has increased over the years, so too has parents' visibility in the decision-making process. Many overzealous parents live vicariously through their offspring, pushing their son or daughter to out-achieve other over-achievers. However, these instances of parental over-involvement have less to do with finding a quality institution for their offspring and more to do with what they can brag about at the next coffee session with their friends. As Robert Samuelson wrote in Newsweek, "What fires parents' fanaticism is their self-serving desire to announce their own success. Many succumb; I did."[1]

This increased competition to get into university has created the false assumption that the greater the demand to get into an institution, the higher the quality must be. Prestigious institutions such as Harvard in the U.S., Oxford in the U.K., and the Indian Institutes of Technology (all ranked in the top 50 universities worldwide[2]), receive many more applications each year than they can accept. Yet, a study conducted by Alan Krueger, an economist from Princeton, and Stacey Berg Dale, from the Mathematica Policy Research, comparing students from highly prestigious and non-prestigious schools, found that students from non-prestigious schools "earned just as much as graduates from higher status schools."[3] Furthermore, in an 827-page evaluation of hundreds of studies of the college experience, Ernest T. Pascarella, the Mary Louise Chair and Professor of Higher Education at the University of Iowa, concluded: "We haven't found any convincing evidence that selectivity or prestige matters."[4]

Despite such findings, perception has outshone reality, creating a booming market for guides to the "best" universities. Research conducted at the Educational Policy Institute in Toronto and covered in the report *A World of Difference: A Global Survey of University League Tables*,[5] confirms what many of us suspected: it's impossible to objectively compare universities. The report concluded that within a national context, there was "no single indicator in common," and "positions of certain institutions in their national rankings are largely a statistical fluke," to say nothing of how they compare when using other countries' ranking indicators.[6]

The report concluded that of all countries, Germany utilized the soundest method of assessment. By simply making the data publicly available, they enabled potential students to investigate and arrive at their own conclusions, based on what they valued in a university. This puts the end users, the students, back in the driver's seat, as they should be. When it comes to picking which university to attend, don't be solely guided by a national ranking index --- meet people from the institution, talk to post graduates, and use your own values to determine which institution is the best fit for you.

Engineering—a wider purview

Once you've chosen a school and been accepted, it's down to business. Your studies will challenge your intellect and open your eyes both to the complexity of the world around you and how much you take such complexity for granted in your everyday life. Your university experience is designed to transition you from a receiver of information, unquestioning of its legitimacy or authenticity, to someone who can argue a point from a particular perspective, educated by facts from sources, and mindful of the differing opinions of your peers. As you learn more about engineering, so your perception of it will change.

At the outset, your view of engineering will necessarily be narrow and focused. At its core, engineering is the application of science; therefore, it should not surprise you to find that much time is devoted to teaching the fundamental ideas of mathematics, physics, and their various sub-groups. At this early stage, you may tend to look at engineering in the following simplified way:

Engineering = f {technical}

where:

technical = (mathematics, physics, statics, dynamics…}

In non-mathematical terms, engineering is a function of technical elements, where these elements are mathematics, physics, statics, dynamics, etc. What you won't realize until much later on, probably after you graduate and enter the workforce, is:

Engineering = f {technical, ethics, communication, teamwork, intuition, creativity, innovation…}

In coming to this realization, you'll begin to appreciate that engineering is much more multi-faceted and non-technical than you thought while at university.

Issues at university

There are many issues at the university that you may not be aware of until you experience them. The following is a non-exhaustive list of issues I experienced first hand, or that I have discovered during my research:

Teaching may not be a lecturer's foremost concern

The goal of most who lecture full time at university is to obtain a permanent position until retirement, otherwise known as tenure. Tenure is gained through the quality of research performed (including the number of peer-reviewed articles published), teaching effectiveness, and the level of prestige with which the individual is regarded within the academic community (external letters of recommendation). While teaching is one of the elements of consideration for tenure, there is a perception that "the emphasis on research and the dollars it brings in eclipses teaching at some schools."[7]

The phrase heard most commonly among those seeking tenure is "publish or perish." For junior professors, this means that a certain number of peer-reviewed articles or papers must be produced within their temporary contract period at the university, or else they do not achieve tenure. Knowing this, it may come as no great surprise why teaching may not be the top priority for many junior lecturers. The less time spent in lectures dealing with undergraduate students, the more time they can devote to research and getting published. This may seem reductionist and harsh, but it's a reality for many academics. In order to secure a permanent position one needs to do X, and X may not include a high priority being placed on teaching outcomes. In an article for the American Society For Engineering Education, Professors Wankat and Oreovicz from Purdue University commented "Until recently, teaching was pretty much ignored by engineering departments at research institutions when it came to tenure. For the most part now, adequate teaching is a minimum requirement, but the decision is still usually based on research and funding."[8]

Until teaching is seen as significant as research, students will continue to feel left out in the cold, seeking assistance for their learning elsewhere or just bumbling along, never really grasping the content of their classes. Those universities that do expressly place focus on teaching are in the minority. John Dodge, Editor-in-Chief of *Design News* commented "Olin, along with Rose Hulman and Harvey Mudd, has a unique approach to engineering education that emphasizes teaching and student engagement. That's not to say other schools don't, but that perception exists."[9]

Marks versus understanding

Many students quickly realize that their university instructors don't necessarily care about their well-being --- different, perhaps, than their experience in high school. This indifference means there is no one constantly looking over their shoulder; now they're on their own. This can spell trouble for students used to structured environments suddenly inundated by lots of work. For many, this lack of supervision equates to total freedom --- but this freedom cuts both ways.

Engineering courses typically involve many hours per week in lectures and laboratories. But now we add in a vibrant university social life. For some, university is a coming of age where, for the first time, they can truly come out of their shells. Just as there's no teacher looking over their shoulder, Mom and Dad are no longer there either.

With the smorgasbord of new experiences and party opportunities, many students pressed for time will invariably use a "divide and conquer" approach. This dictates that when an assignment is given, you get together with a group of friends, pull it apart and leverage off other members of the group. Now there's nothing wrong with working in groups: it enhances understanding when you have to explain a concept to peers, and you're much more open to feedback from your friends that anyone else. However, a real danger emerges when the group doesn't just assist your understanding but rather becomes a substitute for your understanding, something that's relied upon to get you through assignments.

While we usually conclude that good grades connote understanding, is this really the case? It is possible to get good grades on a test, assignment or whole course and yet not really understand what we're supposed to have mastered; conversely, we can get lower marks yet understand more than the grade would indicate.

Looking beyond the marks

Professor James Wilkinson, director of the Derek Bok Center for Teaching and Learning at Harvard University, gave the Menzies Oration at the University of Melbourne, Australia in 2006. Titled, Tests Are No Great Gauge of Learning, Professor Wilkinson made this pertinent comment: "Students are not necessarily learning what they think they are. Just because they pass courses and get a degree does not necessarily guarantee anything except that they are good at taking examinations." He went on to say: "I believe that most faculty at Harvard and elsewhere are genuinely unaware of how little their students are learning."[10] Such comments indicate that we may be fooling ourselves if we think that a certain mark equates to a certain level of understanding.

The rise of cheating in engineering

Getting through the odd assignment or two with the assistance of your peers is one thing, but cheating as a matter of course is something else. You might think it highly unlikely in this day and age of greater scrutiny, but students have figured out increasingly innovative ways of cheating. From the use of cell phones to writing on the inside label of a water bottle, creativity in the method of cheating is only surpassed by its prevalence. Unfortunately, cheating may be on the rise.

In 2004, ABC's 20/20 program, hosted by John Stossel, focused on the issue of cheating at U.S. universities. Everywhere their investigation took them, the numbers of students who had not only cheated, but thought that cheating was acceptable, was on the rise.[11]

Some data suggests that engineering students may be the biggest culprits, but why this is so remains unclear. Professor Trevor Harding, an Associate Professor of Manufacturing Engineering at Kettering University, contributed to a study trying to answer this question. The study, *Perceptions and Attitudes Towards Cheating Among Engineering Undergraduate Students* (PACE), surveyed 650 students from twelve schools in the U.S. and abroad. While they found numerous contributing factors, knowing what they are and the extent of their influence is still unclear. Professor Harding commented: "The data does suggest there is something 'special' about engineering such that these students tend to cheat more than other students do. We are currently in the process of trying to figure out what that is."[12,13] A 2006 study [14] of 5,300 students in the U.S. and Canada found that 54 percent of graduate engineering students admitted to cheating, second only behind graduate business students, who came in at 56 percent. One of the lead authors of the study, Professor Donald McCabe, commented: "Students have reached the point where they're making their own rules."[15] I would suggest that while some responsibility does lie at the feet of the students, modern societal conditioning also deserves some of the blame.

There are indications that what constitutes cheating is changing, and we may be preparing students to cheat at earlier ages. Instant gratification, the "I want it now" attitude, seem much more pervasive than in previous years. Commercial entities exist to satisfy this desire, to do otherwise would be at their own peril and profits. Customers reward (with their business) those companies who can satisfy their needs, and disregard those who don't; businesses have responded. We've entered into an ever-tightening response time between customer demand and service. What other choice is there? Since when is it the commercial world's role to tell the consumer they can't have something? It's the whole point of a free market system and living in a democracy. Can you imagine any business, especially one targeting the teen and twenty-something market, proposing the customer wait a week before downloading mp3s, or suggesting they get parental permission before playing this or that video game?

The question for us is whether this kind of consumer attitude is appropriate in the classroom. Is it reasonable to expect children to understand where a buyer/seller relationship is appropriate and where it is not, such as in school?

At a middle school in Kent, Washington, students are being taught to take tests with somewhat of a different perspective in mind. No longer do they take tests armed with the knowledge and understanding built up by mastery of the subject but instead, during pop quizzes, they go online to search for answers. The motivating factor behind this is to encourage them to feel comfortable on the information superhighway, unlocking the wealth of knowledge residing in cyberspace. In the words of one teacher at the school, "What I'm hoping is that they can find information to help them become better thinkers."[16] While encouraging independent thought is precisely what we want to engender in children, the use of the internet during a test would seem to undermine the purpose of testing -- that is, to test students' fundamental understanding of a subject, not someone else's.

To be clear, I am not against "open-book" exams --- studies on their utility have indicated they "emphasize practical problems and reasoning rather than recall of facts."[17] However, if students are required only to access information from elsewhere, we are implicitly reducing the importance we place on the retention of information. It may be no wonder, then, that students have no qualms about cheating in university, because we may be legitimizing that behavior for them in grade school!

Offenders rationalize cheating with "I'm learning," "the system made me do it," and the idea that if solutions are available out there on the web, how can you expect me not to cheat?[18] While these responses are bad enough, a more sinister problem underlies them, one with serious consequences for organizations hiring these people: if it's okay to cheat in the classroom, how will they behave when they enter the workforce? Compounding the problem, businesses have no way to know if potential employees have cheated at school, due to the university's unwillingness to crack down on offenders. We will discuss some of these issues further on the section concerning ethics.

Grade inflation

Grade inflation is another aspect of the previously mentioned "quality versus quantity" argument among engineering graduates. In 2004, Princeton University, in response to evidence of a widespread culture of grade inflation, reduced the number of A's it handed out to students across all disciplines.[19] Reflected in the 2003 academic year, 48 percent of all students completing undergraduate courses received an A grade. While the goal of the university was to reduce this to 35 percent in engineering, the number has been reduced only to 43 percent. Deflating results in response to evidence is almost as ambiguous as inflating them, and one would hope that the university investigates the root cause. This is not an isolated trend; many Ivy League schools are intently focusing on grading reform.

Grade inflation gives students an inflated perception of their capability. Conversely, it devalues what it means to earn an A and, extended further, what it means to earn a degree. If everyone graduates with outstanding marks, employers must look to graduate degrees as a way of distinguishing who is average from who is outstanding. Even worse, it gives employers a false representation of their potential employee, advertising one product when they are getting something completely different. Professor Brian Manhire from Ohio University commented on his hompage "When college professors give easy grades, we show that we do not respect our students ... [and we] require sacrifices of ... both the exceptional students, who are unable to distinguish themselves because the merely fair students earn grades every bit as good, and the marginal students, who believe they know more than they do because we gift them with grades their work does not earn." [20,21]

Higher degrees

After years of toil, graduation finally arrives, but now you're faced with a decision: to go into the workforce and start your working life or, if your marks are good enough and you have a strong desire to undertake further study, go back for a masters degree, or possibly even a PhD.

You can't avoid further study if you want to stay in the engineering game. It's that simple. If you do not update your skills through continued education, be it formal graduate courses, short courses or self-paced learning, your skill set will erode over time. Even within a short period of time after graduation, some of the things you learned during your education will have been superseded and, in some cases, rendered obsolete. Some of the arguments for and against jumping into a graduate degree program straight after completing undergraduate studies are shown in the table on the following page.

When weighing your options, keep in mind the underlying reasoning behind why you may wish to pursue a graduate degree. I worked for two years after completing my engineering undergraduate degree, and then I entered a masters degree program. I did this because I wished to gain employment in the international space arena, an area I was passionate about. I had a specific tangible reason for investing the time and energy in gaining a higher qualification. My reasons for completing an MBA were also related to my interests and direction of employment.

A reason for NOT pursuing a graduate degree is for its use as a hiding place because you don't know what else to do. People often think doing a masters degree will give them more time to decide what they want to do. Undertaking a further course of study in order to put off a difficult career choice is futile: you'll only have to address the issue again, albeit a year or two into the future.

Everyone's got to jump into the workforce at some point, and it's unlikely that the first job you get is going to be your last. Many opportunities will open up for you once you're in the workforce – consider that before deciding to back out and pursue an advanced degree, if that's not really what you're sure you want to do.

Pros	Cons
Extend your basic knowledge to an advanced level	Expensive unless you have a sponsor (scholarship, employer, etc.)
Recent currency in studying, being a student	May not know exactly which area to do a graduate degree in
Know faculty and can talk to them about which choice of course is best	Opportunity lost in work
Possibly start at a higher wage level when entering workforce with graduate degree	Stigma attached to having a graduate degree, especially if it is on the back of an undergraduate
Makes you stand out from the rest of the crowd of job applicants, in both positive and negative ways	Some practical experience may be required to fully understand advanced coursework

> **Which way to turn?**
>
> At a 2005 IEEE conference, I met John, a power engineer from China who now works in the United States at a laboratory in the Northwest. After finishing his undergraduate degree majoring in power engineering, he wasn't too sure which way to head, so he decided to undertake a PhD. While his PhD is not unrelated to the context of the work he is doing, he admitted that more of his time is spent with non-technical rather than technical issues. "I'm at a crossroads," he commented. "Do I stick to my technical side or do I embrace more than that?"

There are many views concerning the wisdom of pursuing a graduate degree in the workplace. Look at most job advertisements, particularly in the U.S. --- they usually say: "Experience may be substituted for a higher degree." But does having a higher degree really count for three or four years experience on the job? While a postgraduate degree can help in positions that require advanced technical knowledge, what about the vast majority of other technical positions that don't have such a requirement? Is the higher degree really going to place you ahead of someone who has spent the last two years finding the ins and outs of an organization, building rapport with employees and customers, and getting to know the market? For the majority of workplaces outside pure academia or research, an advanced degree won't put you ahead --- in fact, employers may view you as lacking concrete skills that can be used in the workplace, and you may well be treated the same as someone fresh out of university with nothing but an undergraduate degree.

Beware of the "qualification accrual" trap. Some confuse higher education with increases in salary and opportunity. There comes a point beyond which further formal education yields diminishing returns, and your value to potential employers will actually be reduced. I've known people with three or more masters degrees who mistakenly thought these qualifications would yield a top of the line salary. Sadly, this is not the case: three degrees are not necessarily better than one. In some cases, less is more.

Education is all about application and the potential value it adds to the employer's goals. If you are not able to convince employers that your extra qualifications translate to extra value, your degrees will most likely be disregarded. You may only succeed in alienating yourself to an employer by being overqualified or, even worse, inappropriately qualified.

Application of a PhD

I once asked a colleague involved with advanced research for the U.S. Defense Department if he'd used his PhD qualification much since he graduated. "Truth be told," he said, "the only thing my PhD gives me is credibility." During meetings he'd offer a viewpoint or answer a question on a program he had little knowledge of, but people would put more weight behind his response purely because he had his PhD. At the end of the day, a PhD doesn't qualify one in any way to lead a team or handle a major project, yet there is a certain degree of expertise that one builds up, not only in the topic researched but also in general problem solving. As stated by Marcus du Sautoy, a Professor of Mathematics at the University of Oxford, "[For a PhD] you are moving from doing exercises that you know have a solution to questions that no one knows the answer to."[22] When viewed in this light, it would seem that for engineers, a PhD would have great appeal to their problem solving role.

The MBA

As engineers have become more multi-disciplinary in recent years, there has been much made of the utility of an MBA. Many think that with one of these tucked under their belt they'll be able to demand a higher salary, and it's pretty hard not to be wooed by potential six-figure earnings right after graduation. Sadly, however, there's more marketing hype than substance in these claims. Business schools don't tell you that you need a number of years in the workforce to properly utilize learning contained in an MBA, hence the reason why some schools will only admit candidates with at least a few years experience.

It's worth noting that for engineers, an MBA is not to be undertaken lightly. As engineers, we are used to the scientific method, logic, clear and concise answers of right and wrong, yes and no. In an MBA, you must be prepared to suspend this kind of thinking and deal in the realm of opinion, shades of correctness, and lateral thinking. It can be much more demanding than engineering because there are no universally correct answers, just those arguments that are well-supported versus those that are not.

Many big companies have hired new MBA graduates only to be disappointed by their lack of tangible skills and their one-dimensional focus on theory. This typically results from selecting candidates based on which business school they attended, not on what skills they have accrued.

An employer's perspective on the MBA

An article for the Fall 2004 edition of *Strategy + Business* magazine relayed the results of a survey of over 100 executives from more than 20 countries. The author referred to the poignant comments of one executive: "He would prefer to recruit a candidate with a PhD in Russian literature than one with an MBA. He thought knowledge of Dostoyevsky denoted someone with curiosity and a learning attitude."[23]

This highlights the filter of scepticism through which some business executives view those who possess credentials without the genuine interest and passion to support it. From a study in Europe by Human Resource consultants Cubiks: "Fewer than one in ten (eight percent) believe that academic qualifications are always a reliable indicator of how a candidate will perform in a role."[24] At the end of the day, lectures and book learning can only imbue an MBA graduate with a certain range of skills. For them to truly understand what they are taught in class, an experiential component is required.

University and workplace—bridging the gap

The discussion on university, industry and their respective roles in educating tomorrow's engineers is one that has gone on for quite some time and resulted in two emerging schools of thought.

The first, what I call the "blue sky" argument, states that university is a place for exploration, a place for students to learn the fundamentals of engineering unencumbered by its relevance to the workplace. Proponents argue that if you are to truly create independent thinkers who can generate innovative and creative solutions to modern problems, you need to provide an environment where they can explore. Detractors say universities run in this fashion merely create ivory towers of intellectualism isolated from the real world.

The second camp argues that time spent at university is a means to an end. What is learned in class must be relevant and connected to how engineering is applied in the real world. Proponents argue that real world engineering education has for too long been picked up by industry after graduates enter the workforce. Detractors comment that such a focus on the workforce turns universities into factories that turn out commodities and do not teach students to be independent thinkers and learners.

Co-op Programs

There is merit to both arguments, and ideally university would be a place that tries to instil in students an ability for independent thought and critical thinking as well as equipping them with the skills they'll need in the workplace. One tool that potentially meets these needs is the Co-Operative or "Co-op" program.

The Co-op program is more than the usual requirement for students to complete some form of engineering work experience during their engineering degree. Participating students either alternate between full time work and study from one year to the next, or combine working part time with studying part time. The work is paid, and by graduation the student has accrued a sizeable amount of industry experience, something full time students cannot attain during summer break work experience. Co-op students may take a longer period of time to complete their course, since it is spread out more than studying full time.

I see the intent of the Co-op program as overwhelmingly beneficial: the student is afforded the opportunity to see the application of their studies in the real world early on in their careers. They also receive the opportunity to form strong industry links, usually going straight into full time work with the employer they co-oped with.

> **Co-op program at Georgia Tech**
>
> In 2006, the Georgia Institute of Technology (Georgia Tech) Co-op program celebrated its 100th year in existence, and there was plenty to celebrate. Debbie Pearson, assistant director of Cooperative Education at Georgia Tech, pointed out that "the GPA of co-ops is consistently higher than that of non-co-ops."[25] She also noted many other advantages, such as lower burn out rate of co-op students over their full time counterparts, the effect on retention of engineering students through to completion, and anecdotal evidence that the salaries of those graduates who were co-ops is higher than those who were not.

Professional Associations

Another way of engendering industry understanding while still at university is through the professional association. Typically, professional associations focus either on different industry clusters or specific engineering disciplines; the Institute for Electrical and Electronics Engineers (IEEE), for example, focuses on the disciplines of Electrical and Electronic Engineering, while the American Institute for Aeronautics and Astronautics (AIAA) deals with the industry cluster of Aerospace.

Currently, many professional associations are struggling to maintain relevance. Many have seen their membership numbers slow in growth or stagnate, partly due to the increased scrutiny prospective members place on paying high fees: the "what do I get for my money" attitude. Mr. John Vig, an IEEE Fellow and past President, had this to say: "[I] find it increasingly difficult to tell people about the tangible benefits. Therefore, [I] emphasize the intangible value of being a member."[26] And while giving back to the profession is something that may appeal to those who are somewhat along in their career, how can you strike a chord with those potential members who haven't been in the profession long enough to feel the pull of civic duty? This has led to the development of active content such as IEEE TV, which hopes to provide members with more tangible returns for their membership dollars, representing an information push rather than pull.

There is another issue related to professional associations. In the past, engineering was represented quite well by various disciplinary streams --- civil, mechanical, electrical, etc. Engineering is now far more multi-disciplinary, encompassing all previous disciplines plus others not easily categorized; therefore, associations are struggling to keep pace. Many potential members do not see themselves represented in the current frameworks, and without some kind of binding legal requirement or financial incentive for membership, they either opt out or become part of the passive membership that joins only because their workplace pays for it.

Chapter 2, If you only have 5 minutes...

- Parental involvement in the university decision appears more prevalent now (or at least reported on more) than in previous years.
- The correlation between demand and quality has become blurred, but it is not necessarily the case that those institutions that experience higher demand necessarily provide a better quality of education.
- University guides that compare institutions across the nation may not offer a fair, balanced or even reasonable comparison – ultimately, base your decision on what matters to you.
- Lecturers seeking tenure may be more focused on research publications than course instruction – being mindful of this will help you understand the difference between them and your teachers in high school.
- University offers ultimate freedom – the freedom to do well but also the freedom to fail.
- Be mindful of "grade creep" which reflects a situation whereby high grades do not automatically correlate to high levels of understanding.
- Data suggests that cheating among engineering students is more prevalent than most other disciplines, though the motivation as to why this is so is unclear.
- Modern societal expectations and a greater awareness of the business contract have lead some to view the education transaction in a similar way.
- Grade inflation serves to devalue the status of educational credentials (degrees and certificates) and means that employers push the credential bar higher in an effort to distinguish among applicants.
- Pursuing a higher degree, such as PhD or MBA, will not automatically translate into a higher salary.
- Ask yourself what extra value the education you're pursuing will have for a potential employer.
- Co-op programs represent a good way of obtaining real world experience while still at university, though the drawback is that it may take longer to complete.
- Professional associations are challenged today by trying to create member value, with many doing this through the development of active online content.

Actions to take away

Some things you can do to act on the information presented here:

In deciding on which university to attend:

1. **Write** a list of things that are important to you (want to stay regional/move away, sports, hobbies...).
2. **Discuss** these with your parents, friends and/or other family members.
3. **Listen** to their input, but remember it is ultimately your decision to make.
4. **Research** the different universities offering the courses you would like to take and score them against the list you created in step 1 above.
5. **Evaluate** the pros and cons (for yourself) of doing a co-op rather than straight through degree.

If you are trying to decide on whether to pursue graduate studies or not:

1. **Ask** yourself the following questions by writing a list:
   a. Why do I want to study this particular area?
   b. What will this extra credential give me (credibility, satisfaction, job prospects...)?
   c. What extra value will/could this add to my employer (if employed)?

3. The Working World

> Today's workplace has changed from that of years past. In this chapter, we'll try to address how and why it has changed. Beginning at the macro level, the perpetual lament that "there aren't enough engineers" may not be entirely true given actual data. At the micro level, the employer/employee dynamic surrounding the social contract has changed, challenging the familiar "job for life" paradigm. Similarly, career models are under assault, changing from a linear form of progression to one resembling a spiral. Our time horizons are much shorter now, for both people and organizations, and this has given rise to the "just in time" job. As organizations continue to face the challenge of retaining young engineers, the much-touted multi-tasking ability is shrouded in hyperbole – we'll seek to dispel some of the myths here. Finally, we'll cover some aspects the new workforce may be looking for in their jobs.

Too many jobs, not enough appropriately skilled workers

During the first decade of the 21st century, the birth rate in many developed countries has decreased due to a number of factors: the increased proportion of women opting for career over family, more career choices in general, and greater individual freedoms. Advances in modern medicine continue to extend life expectancy, so the average age of the population in developed countries is rising. These factors, along with the impending retirement of baby boomers, and a well-documented inability to draw more students into engineering,[1] contribute to the perception that there are too few workers entering engineering-related fields --- too few engineers to replace those going into retirement, or to fill new jobs in industries yet to be realized (such as nano-technology, etc.).

I say "perception" because, despite the popular belief that there are not enough engineers in the pipeline to satisfy future demand, a number of recent studies and reports invalidate this claim. In the report, *Into the Eye of the Storm: Assessing the Evidence on Science and Engineering Education, Quality and Workforce Demand*, authors Lindsay Lowell and Hal Salzman state that "managers in engineering and technology firms do not claim a shortage of applicants nor do they complain of applicants with poor math and science skills or education. They do often note difficulty in finding workers with sufficient experience, specific technical skills, or a sufficient number of 'brilliant' workers in the pool."[2] Furthermore, the claim that the U.S. must graduate more engineers because it has been overtaken by India and China misrepresents the situation. While it is true that these countries graduate more engineers, such a myopic focus on numbers suggests an oversimplification of the situation.

First, the term "engineer" is not interpreted homogeneously, as documented in work done by Professor Gary Gereffi et al; "Varying conceptions of the engineering profession exist not only between countries, but even within them."[3] Secondly, the reported figures do not take into account the *quality* of those graduates as I first mentioned in chapter 1. "80.7 percent of U.S. engineers were employable while only 10 percent of Chinese engineers and 25 percent of Indian engineers were similarly employable."[4]

This suggests that the debate over the sufficiency of engineering resources is multi-faceted and requires a view encompassing more that just the supply side.

The new employer/employee relationship

Employers who were once in a position to say "take it or leave it" to employees find themselves in a more vulnerable position today. Much of their work can no longer be done by just anybody, but instead requires those with specific niche skill sets, particularly in the case of engineering. And while declining birth rates create the perception that there will not be enough people to replace those retiring, the main problem isn't a lack of people (the supply side argument), but a lack of the right skills. It is important to understand this distinction because, while those who cannot find work may very well have skills, they're incongruous to the workplace at present and will become more so in the future.

Employees possessing skills demanded by the new economy are increasingly aware of their more valued status and emboldened by greater self-determination. The new contract between employee and employer is often: "I'll work hard for you while I'm here, and in return you'll provide me with opportunities to train and learn, and a level of advancement and progression." The employee is fully aware there is no guarantee of long term employment with the company, making continued training and skills accrual so necessary.

Social responsibilities

Social responsibilities are part of the benefits package offered to salaried employees. It usually includes medical and dental insurance, 401(k) and annual holiday entitlements. Employers are beginning to feel pressure on this front. Providing full benefits like healthcare is a huge financial burden. Examples like GM show how these costs can translate into products more expensive than their competitors. Such legacy overheads were directly responsible for changing GM from an investment grade to a junk grade stock in 2006.[5]

Instead of a cold turkey approach, employers are trying to transition some of this burden to the employee by "increasing employee contributions, offering a wide range of voluntary benefits and introducing more flexible plan designs."[6] Despite the ever-rising costs of medical insurance in countries like the U.S., many corporations "felt that they needed to offer and subsidize a wide range of benefits as part of their recruitment and retention strategy."[7] Benefits represent a source of differentiation when trying to lure talent.

Employers are also placing greater scrutiny on training. In light of actual or potential high employee turnover, employers commonly argue: "Why should I provide training when they're going to leave five minutes later to get a better paying position somewhere else?" As stated earlier, employers are in somewhat of a powerless position here. They either (1) don't provide training, and don't attract the talent they need, or (2) provide training, and watch new employees dissatisfied with the workplace use it as a springboard. The outcomes of this apparent choice between a bad and worse situation hint at a potential resolution: the employer must insure the employee's experience in that organization is so positive they won't want to leave, or if they do, they'll recommend working there to others. We'll discuss this further in the section called *Rhetoric versus reality*.

The employer should be fully prepared for an employee to change jobs at any time. The saving grace for the employer is that similarly disaffected employees will be leaving other companies and approaching his. This is where their return on investment comes in. The employer is now an enabler of the employee in establishing their career path rather than being the dictator as they once were.

The employer should provide an environment conducive to receiving and transitioning people to and from the organization. Once they see themselves in this way, losing employees won't be seen as a problem, but rather an asset. If the employee received excellent training and a few good years with the company, they're more likely to speak positively about them at their next job. Get enough people talking about your company in this way and you won't have to advertise for the best and brightest talent, they'll be lining up at your front door. Nothing sells potential employees better than referrals from a satisfied workforce.

This approach is not without its drawbacks, and it does require a more harmonious relationship with competitors than probably now exists. You see this employee turnover in industries such as aerospace, where an employee may, for example, leave Boeing and then go to work for Northrop Grumman. The upside to this is that much of the culture and work is very similar; the downside can be exactly the same: that much of the culture and work is similar.

The new environment

Our notions of work are under assault in the modern era. Now, more than ever before, we are able to predetermine the path we follow. Unfortunately, the multitude of options available can be paralysing. The traditional model moved you into a career previously worn in by others, but those entering the workforce today can no longer be comforted by the knowledge that there is a well-trodden career path to be followed from one level to the next. As an engineer in today's workforce, your jobs are likely to be uncoupled and disjointed; you will likely move to the top of one ladder only to change employers or functions and find yourself starting out at the bottom of another.

Year	Age	Work	Travel	Formal Education
1995	20	Student	Australia (W. Coast)	Engineering undergrad (2nd, 3rd and 4th year)
1996	21		Korea	
1997	22	Air Force	Australia (W. Coast)	
1998	23		Australia (E. Coast)	
1999	24		Europe, US	MSc (full time)
2000	25			
2001	26			
2002	27	Consulting (individual)	Australia, Europe, US	MBA (part time)
2003	28			
2004	29	Salaried worker		
2005	30	Consulting	US	

This is demonstrated by using my own career to age 30 as an example. This path has continued beyond 30; currently I work for a company while pursuing a PhD.

As seen in the above table, and in your own experience to a

greater or lesser extent, our careers now follow a spiral path rather than the traditional linear or stair step model. And so, just as the length of time spent in any one job has decreased along with our horizons, so too has the timeframe companies place on getting the technical talent they need.

Just in time jobs

The reduced time horizon most companies experience has created the "just in time" job. I first heard this term used by Ed Perkins, from the IEEE Portland section, at an IEEE conference in 2005. It accurately describes the way these jobs are created.

Essentially, a "just in time" job is created for a specific organizational task or requirement. It may be to create a certain piece of code, develop an application, create a particular circuit design, or lead a technical team. The employer is after someone who has done the task a number of times before, and will therefore be able to come in without need for training or extra knowledge: hence the term "just in time." Once the task is completed, the requirement dissolves and assuming there are no further needs, the short term employee (usually classified as a contractor or consultant) is let go.

In this mode of employment an employer needs an experienced candidate, having no use for those straight out of university, in their first few years after graduation, or even those who have worked for many years. The key here is skill set. Most employers will be unwilling to train an incumbent to get "up to speed" as they want the task performed now --- the employee must "hit the ground running."

In my opinion, this sets an organization up for trouble. First, while it is understandable that, pressured to bring projects in on target and become more competitive, there is a constant need to "trim the fat." But when you continue to call on the same experienced people, and not take the time to transfer knowledge and train new workers, you're operating a closed shop, and when those resources you solely rely on retire, you have no qualified staff to draw on. We are currently seeing organizations that have leaned heavily on certain individuals scrambling to transfer their knowledge and succession plan before they retire.

Secondly, handing such specialized tasks off to specific individuals outside the organization, and requiring them to do only that task, does little to develop new employees within the organization. The outsider's skills are not retained, and once their task is completed the organization will need to look outside again the next time such a requirement emerges.

Thirdly, operating in this way sets up a discrete mode that moves forward in fits and starts. There is no continuity when different people are brought in as different needs arise. Today's requirement may be gone tomorrow, in a week or in a month. This constant turnover is likely to make the organization feel unsettled to permanent employees, who may be continually going over old ground and explaining ideas to new employees.

Finally, because the contractor only does work in the area of their skill set, they are never encouraged to develop other skills and/or expertise. This means they are less likely to understand points of view of those coming from different backgrounds; they may not speak the same language, or might not be aware of the concepts they are trying to explain.

So while I can appreciate the need for "just in time" employment, it does nothing to create the balanced engineers a company would wish to hire for the long term. Instead it creates a series of technical specialists who have all their eggs in one basket, hoping there will be a constant need for their specific skill set. If that need dissolves, they will become irrelevant to the market. While there will undoubtedly be continued short term needs for contract staff, the real value added work needed by organizations requires a more strategic appreciation.

How can those already on the inside of an organization fulfil these "just in time" needs? One solution is to encourage students and young graduates to enroll in short courses specifically matched to employers' requirements. These courses could be aligned with a particular professional society so, upon completion, the student would be able to show an employer a certificate with a professional society's seal, stating that John Smith had completed a particular course in, e.g., VLSI design. You might say: isn't that what going to university is all about, and isn't that why students take courses in various engineering subjects? Well, yes, but during university students get basics and fundamental grounding in various subjects. Their learning is not directed or focused in a particular way, which is what these short courses would do; they would be applied or developed with a particular purpose in mind. (I first heard this idea from a colleague of mine from Texas, David Walker, so I must give him credit for coming up with this.)

Rhetoric versus reality

So the question arises: why do organizations continue to struggle to get not only young engineers but new young employees? Is there some reason other than a lack of supply that has not yet been identified? I emphatically say yes there is.

> **Toxic companies**
>
> Professor Jeffrey Pfeffer, who lectures on Organizational Behavior at the Graduate School of Business at Stanford University, uses the term "Toxic Companies" for organizations who complain that they can't get talent, yet treat employees as commodities. His view is that most companies "get exactly what they deserve," because they "drive people away with their insanely long hours, limited career mobility and work practices that treat all employees like potential litigants."[8] As a result, people jump from one organization to another, hoping they'll find some place that's different from the others.

Various trade magazines and committee reports continually remind young engineers of the need to be more multi-disciplinary, broader in outlook and more flexible. Even I do it in this book, and I do genuinely believe in the need to posses these traits. The disconnect comes when you look at what is actually happening in the workplace: such flexibility, multi-disciplinary nature and ability to adapt and handle change are rarely mentioned or practiced by companies who preach it. They are typically the first things to go when it becomes apparent that maintaining such a culture takes discipline from the top-down, something few organizations consistently maintain. At present, many organizations believe *saying* they having such cultures makes it so, the equivalent of wishing a boulder up a hill. No matter how much you want it to, it won't happen without effort. A recent study of 85,000 people over four continents bears this out. It found that "only one in seven employees worldwide are fully engaged in their jobs and willing to go the extra mile for their companies."[9]

Despite espousing the traits of flexibility, multi-disciplinarity and adaptation to change, my own experience in trying to obtain a job in Silicon Valley bears this out. I found it increasingly frustrating to run into the arcane argument that specialization was everything, and outsiders couldn't possibly penetrate this or that industry due to its intricacies. Even pointing out similar environments, showing an understanding of issues confronting the industry, or making suggestions on how to solve their most pressing problems was often met with a dismissive attitude. Herein lies the difference between rhetoric and reality.

While we think a strategic appreciation and a top-down focus on the issues is what employers want, in actuality they hire based on job requirements, developed in a bottom-up fashion with minimal thought put into what it is they are really after. These job requirements are not developed with a focus on the issues and problems likely to come down the line -- they are focused purely on the short term, and the faster the organization's cycle time the shorter it's likely to be. You need a software program developed in six months? Then let's hire a heap of java programmers immediately. This gives rise to the "just in time" jobs phenomenon described earlier.

Looking at the U.S. engineering market, particularly the tech sector, from an outsider's perspective (as I do), those within are highly specialized. In essence, they "look at things through a soda straw." Such a narrow focus creates huge problems within their organization because there tends to be no cohesion or group focus, messages and intent get scrambled, and no one seems to know how to step above it all and look at the bigger picture.

In an attempt to work with a greater sense of unity, organizational restructures try to find the best way to orient. However, this effort commonly results in employees not knowing who they're reporting to this week, and in changes that never affect how things happen at the working level. Tasks seem reactionary, relating to solving short-term crises that need to be done "as soon as possible." And so, despite the rhetoric, many employers still do not understand what they want in an employee, or what it takes for an employee to be successful. It's for this reason the label "toxic employer" continues to fit many organizations rather well.

Generation Multi-task

To many baby boomers, the way Gen Y uses their time is a foreign concept. A new tag, Gen 'M' where M stands not for Millenial but "Multi-task," more accurately reflects one way this demographic has come to be defined. Every employer wants an employee to manage more than one task at a time. Otherwise, tasks would be completed in a serial fashion, with no adherence to priority or urgency -- essentially the slowest way to work. A constant need for movement, stimulation and dynamism by Gen M characterizes them at the other extreme.

A 2006 *Time Magazine* article exposed a growing problem many fear about today's younger generation: they are too plugged in --- and such connectedness may come at the expense of face to face interaction.[10] While plugging in affords Gen M the ability to connect with far more people than ever before, what quality of relationship are they able to develop in cyberspace, and can this be substituted for face to face social interaction?

This issue is now being discussed at universities as they try to move with the times and adopt technology to facilitate the learning process. And while the introduction of class notes and course information on the web when and where people want it is a good thing, questions have been raised over how much benefit technology's use in the learning process provides.

Take for example the increased use of laptops in lectures. In many engineering classes, students no longer go flat out trying to write down lecture notes from the board with pen and paper; now they type them in as the lecturer is speaking, interact in online chat-rooms about the lecture content, or if the lecture notes are downloadable -- better yet, their time is free to just sit there and absorb the lecture. On the surface this would seem ideal; one of my biggest gripes from my undergraduate years (especially in mathematics classes) was that I spent the whole period frantically trying to get everything down that was on the board, leaving no time to absorb what the lecturer was saying. That all came later (hopefully). The problem emerges when technology competes with the lecturer for the student's attention.

Multi-tasking in the lecture theatre

At the University of Houston (UoH), Computer Systems Professor Dennis Adams comments: "You can be in the front of the classroom and your hair could catch on fire and they'll never see it because their eyes are glued to the 14 inch screen at the end of their nose."[11] Aside from the irony of a computer systems professor bemoaning the use of computers, his point stands. Students who engage in writing emails and using chat rooms cannot be engaged with the class, and have little chance of taking in what is being said -- and there is quantifiable evidence to back this up.

Research conducted at the University of Michigan challenges the notion of being able to do several things well at the same time. "If a teenager is trying to have a conversation on an email chat line while doing algebra, he (or she) will suffer a decrease in efficiency, compared to if he just thought about algebra until he was done. People may think otherwise but it's a myth."[12]

Relating this to the study of engineering, while such multi-tasking may help Gen Y's divide and conquer the academic syllabus of an engineering degree, it won't aid them in actually understanding what's being skimmed through. It is breadth at the expense of depth with no actual impact on understanding, and this can be a big problem when we're talking about the complex issues engineers deal with that require layers of understanding.

The big crunch comes when entering the workplace; those missed opportunities to develop communication skills through basic person to person interaction in the early years become obvious, and can be difficult to recover from professionally. This is why it is so important that if you are someone who has a propensity towards electronic communication, make sure you take time out from instant messaging a person and interact with them *face to face* if at all possible.

So just what does Gen Y want?

So just what is it that these new young professionals want from their career? Some of the prevalent aspects include:

A chance

Those children born after 1980, commonly referred to as the Millenial Generation or Generation Y, are opposite in many respects to the baby boomers, those born between 1945 and 1964. The Millennials, and Gen X-ers to a lesser extent, share nothing of the expectations the baby boomers had for life long careers. They also do not buy into the bureaucracy of large companies, or respect positional authority as a valid method for a boss getting what he wants.

In a scene from *Reality Bites,* a movie about four Gen X-ers graduating college and facing the turmoil of what comes next, Winona Ryder's father, a baby boomer, turns to her and says: "You know the problem with your generation, you don't have any work ethic." While this comment is about Gen X-ers, it applies to Gen Y's as well and highlights one of the problems facing restructuring the workforce to meet the modern worker -- it's still being run by those baby boomers who worked long hours and now expect Gen X-ers and Gen Y's to do the same. It confounds many baby boomers that the pursuant generations do not want or expect to work inordinately long hours.

Having seen their parents place work above everything else, and the shortcomings such a lifestyle brings, Gen X and Gen Y are looking to make their work fit their lifestyle, not the other way around. They want the flexibility to work when they are at their most creative, and choose those areas that are of greatest interest to them. They don't want to be dictated to or pigeon-holed, and opportunities for continued professional development are paramount.

While Millenials think of this desire for constant movement and the opportunity for advancement as ambition, baby boomers see it as wanting to shortcut the system by not "paying their dues." There is some truth to both sides of the argument. On the Gen Y side, it is perfectly legitimate to expect and receive professional advancement based on the caliber of one's performance. What is unreasonable, however, is the unrealistic expectation that advancement up the ranks is a matter of due course, with little consideration as to how much time and skill it might take to attain these goals.

At a technical conference I attended, a baby boomer recounted his experience of undertaking a PhD with 20-somethings. The prevailing message he got from them was that television programs such as "The Apprentice" have made popular the notion that anyone could be plucked from obscurity and, within a matter of weeks, be qualified for a shortcut route to advancement. The need to accumulate experience at each level before taking on bigger responsibilities at the next level seems to have been lost. It is one thing to understand the theory, it is not such a linear progression to truly master an area of expertise. While this takes time, Gen Y's may perceive this necessary step as a waste of time.

A voice

Young engineers are not sitting back waiting to be catered to. Instead of passively sitting in organizations until they are considered senior enough, young engineers all over the world are pro-actively talking among themselves and with senior colleagues about what they want, resulting in special breakout sessions at conferences and sessions wholly-dedicated to providing a voice to young engineers. Examples include the International Young Professionals conference, held for the first time in 2005, and the selection of young technical professionals by the European Space Agency (ESA) to attend the International Astronautical Congress (IAC) each year. These events provide young professionals (not just young engineers) the opportunity to discuss topics specific to themselves, as well as interact with senior members of their own and other organizations, people whom they might not otherwise get a chance to speak with.

Mentoring

One of the most important things new young employees want when entering an organization is to be mentored and shown the ropes. These mentors can be both specific to their new job as well as removed from their hierarchy, affording them the opportunity to talk about larger issues. For mentoring to be successful, however, there have to be some ground rules.

Mentoring must be conducted in an open and non-judgmental way. Discussing issues you want independent feedback on only to have them trivialized, derided or made fun of, does not good mentoring make. In these cases, the mentee will be far less likely to bring issues to the mentor, and will likely seek out alternative avenues for assistance --- in effect, find themselves another mentor.

Mentors should be chosen, not assigned: The relationship between mentor and mentee cannot be an artificial one, and it has to develop from the ground up. A good rapport needs to be developed before trust can emerge. Without a foundation of rapport and trust, there is no chance for a mentoring relationship to survive.

An informal approach is preferable to one that is structured or mandated: meeting every week or two to discuss topics the mentee wants to address, in essence a push or bottom-up system, is preferable to meeting at a pre-determined time and place to discuss issues as dictated in a formal mentoring policy, in essence a top-down or pull system.

A mentor must guide: In most cases, mentees go to their mentor for advice and guidance on a certain situation. In order for the mentee to learn and grow, they need to receive encouragement and support to arrive at answers on their own. Giving mentees answers lays the foundation for a relationship of co-dependency, inhibiting their ability to develop their own cadre of skills and confidence in dealing with a variety of situations. Because of this, the mentor should always seek to provide guidance or facilitate the mentee's own progress, and take care to never pull them along.

There must be an agreed upon level of commitment: it is true that you only get out of the relationship what you put into it. Both sides must be sure they are committed to putting in the effort, otherwise one side will ultimately come away disappointed. I once entered into a mentoring relationship with a senior figure in academia only to be annoyed and frustrated by the lack of commitment the other party had to the relationship. It was ironic that he submitted his name to be part of a mentoring program with a professional society!

While it is traditionally perceived that the mentee is going to be the recipient of knowledge, they usually have a lot more to offer the mentor than they realize. A mentor can get just as much, if not more, out of the relationship, picking up new knowledge and perspectives especially when mentoring younger colleagues. This helps mentors keep in touch with issues and problems being faced by those at levels far removed from them in everyday settings, outside of their usual formal channels.

I have had a number of good mentors so far in my career. They weren't necessarily those who guided me on specific work issues, but rather people I used as independent sounding boards for advice on bigger, more strategic career issues. I sourced them based on pre-existing relationships (usually built up through professional societies), my professional respect for them, my sense of their good judgement, and finally a feeling that I could talk to them and not be judged on what I was going to say, or not be told what I should do. On the flip side, I have also sought the advice of those who turned out to be not as good as I predicted, who tried to dictate what I should do in certain situations. Suffice it to say, I thanked them for their suggestions but did not seek their counsel again.

Opportunity for Overseas work

Consistent with their more globally-focused lifestyles, many Gen Y's want to see the world and work in foreign lands. In his book, *The Flight of the Creative Class*, Richard Florida argues that the employment field is now truly global. He writes that talent has started to move around the world, as those possessing the skills look for areas where like-minded people are already established to set themselves up. While there are still some impediments to the free flow of human capital (e.g., work visas), those with the right kind of skills would ideally be able to work anywhere in the world they wish. It would be an employee's market, as the gradual aging of the developed world would toughen the fight for talent among companies, offering increased incentives designed to lure the best new employees.

Fed up with it all—starting at the top

As a direct result of their frustrations in trying to get ahead in today's organizations, some young people have taken the bold step of bypassing them altogether and starting at the top. "For the most ambitious young people, the corporate ladder is obsolete," explains Paul Graham, a partner at Y Combinator, a venture capital firm that deals primarily with start-ups. While most go to start-ups after becoming disillusioned by a few years in the workforce, others bypass it altogether and go straight to the start-up phase.

Take the case of Alex Ohanian and Steve Huffman who, at 22, started their own company while still students at the University of Virginia. Five months later, they were tempted by a multi-million dollar buyout offer from Google. While they turned down the offer, at 22 they are charting their own destiny and, in their own words, "doing the entrepreneurial thing."[12]

> **Bypassing the career ladder**
>
> Michael Brett, also 22, graduated with an engineering degree in Aerospace Avionics and spent two years in a traditional company before deciding to join with other like-minded individuals and follow the start-up route. When asked why he left his previous employer, he commented: "It was a good starting point for my career but at this stage, the start-up culture is a lot more appealing." Now, six months after joining his new employer, he is the Chief Engineer for Snow Sports Interactive, a technology start-up focusing on enhancing the experience of skiers and snowboarders. "I'm learning so much in this role and the satisfaction of having your company grow because of your own actions is second to none."

From my own experience, going out on your own is a baptism of fire, and by that I mean it's sink or swim: you either learn from your mistakes and improve, or you are engulfed by them and don't survive. You learn so much in such a short space of time that the end result is almost irrelevant. Of course, it's nice to be able to build a company for a profit, but it's the process that really matters. Learning the intricacies behind cash flow, customer satisfaction and product development could take years through the normal channels of being in a company, especially as an engineer. You develop highly employable skills that will stand you in good stead should you ever want to re-enter the salaried workforce.

And it's not just relevant to those seeking profitability. There are many others charting their own course through the non-profit and volunteer communities. As mentioned previously, young people today want work to fit their lifestyle, and this means they identify much more with social causes than previous generations. Organizations like "Engineers Without Borders," the "International Young Professionals Foundation," and the response by young engineers wanting to contribute to natural disaster relief for events like the Asian Tsunami and Hurricane Katrina, show that perhaps they are not as inwardly focused as they are commonly portrayed in the media. Involvement in these causes develops leadership, teamwork, communication and cross-cultural understanding, again skills they may not have the opportunity to develop in their regular workplace. Also, considering the well-being of people other than themselves helps to create maturity and understanding.

Chapter 3, if you only have 5 minutes...

- The perception that the U.S. is not graduating enough engineers is an argument not supported by the data, indicating that the "shortage" should be more appropriately referred to as a skills and experience mismatch.
- While China and India may graduate more engineers than the U.S., the figures do not take into account the quality of graduates, nor is the definition of "engineer" used to compare the figures the same.
- The workforce dynamic between employer and employee has changed from one of dependency (a parent-child relationship) to a more evenly balanced one of mutual benefit.
- In the U.S., the continued social contract that employers have shouldered in the past will move to one of more joint responsibility between the employer and employee.
- Training and skills enhancement opportunities represent an increased distinguishing factor as to whether new employees will leave or stay with an organization.
- Employers should be open to providing an environment that is comfortable with transition, as this will ensure them being an employer of choice.
- Career paths no longer follow the stair step model, but rather a spiral model whereby new career episodes build on previous areas worked but with greater responsibility or scope.
- "Just in time" jobs are a method for companies to meet short term needs, but unless attention is paid to transferring knowledge into the organization, it can be lost when short-term contractors move on.
- Short courses focused on specific employer needs represent an ideal way for employees to tailor their skills.
- Despite the perception that multi-tasking is a catalyst to accomplishing more, data suggests that it actually leads to a decrease in efficiency.
- The introduction of technology to the classroom has led to a competition between lecturer and computer as opposed to the envisaged scenario whereby the technology is a catalyst for the learning experience.

- The perspectives of Gen Y, Gen X and baby boomers on work and its place within their lives have been influenced by the environment in which they've developed. This is a reason why these three generations can seem different in their outlook.

- Advancement to higher levels of responsibility within an organization involves both a learning component (on the job and academic or book learning) and an experiential component. It is the experiential component that sometimes cannot be shortcut, requiring time and patience to absorb and observe how things are done.

- Young Engineers entering organizations are seeking to have their voices heard within the organization, and many professional conferences and societies now specifically cater to this.

- Young engineers are hungry for mentoring, as a way to sound out their ideas and seek guidance on their career path.

- Mentoring is a two-way path, with both the mentor and mentee ideally gaining from the exchange.

- Many young engineers want overseas work experience, given their comfort level with being in the global environment.

- The entrepreneurial route can offer an accelerated path to dealing with a variety of technical and non-technical issues young engineers may not otherwise encounter at such an early stage. Things such as dealing with customers, deadlines, cash flow and contracts are but a few of these issues.

Actions to take away

Some things you can do to act on the information presented here:

Employer Relevant skills

1. **Write** a list of the skills that you think your employer sees as relevant to the type of work they do.
2. **Validate** this list with someone at your organization who you think would be able to know with some authority, e.g., Human Resources, training co-ordinator, immediate boss.
3. **Ask** if there may be some opportunities for you to learn some of these skills, either through self-learning (on your own time) or employee training (on the clock).

Career episodes

1. **List** the different career episodes you have had so far.
2. **Think** about what tasks or actions you have taken in each of these episodes and write them down.
3. **Identify** common tasks or actions you have conducted in each of the career episodes. What can you say about the scope or size of the common tasks each time it has successively been done? What can you point to in your experience that enabled you to take on the common tasks, but with increased responsibility and/or scope?

Multi-tasking experiment

1. **Get** a stopwatch or some other timepiece.
2. **Open up** two of your favorite social networking applications (one being your myspace or Facebook page, the other being Instant Messenger or some other real time chat).
3. **Open up** the online page of any newspaper.
4. **Choose** a short article of several paragraphs in length.
5. **Start** your stopwatch or timepiece and begin to read the article.
6. **At the 1 minute mark** (that is, you've been reading for one minute) take note of where you're up to and move to your facebook page and start updating it, sending out friend requests, etc.
7. **At the 2 minute mark** (that is, you've been on facebook or another networking site for 1 minute) move to your Instant Messenger or real time chat and start an online chat session.

8. **At the 3 minute mark** (that is, you've been on the chat session for 1 minute) move back to where you left off in the article (don't go and recap what you've read) and continue reading for another minute.
9. **Repeat** steps 6 – 8 until you are at the end of the article and write down as much as you remember from it.
10. **After a period of time** (1 day or so) sit down and read the article from start to finish without distraction.
11. **Write down** as much as you can remember from it.
12. **Compare** what you wrote in the first instance and the second instance. Is there anything you notice? What does this tell you?

Overseas work

1. **Make a list** of those countries that you would like to work in.
2. **Research** the national engineering bodies for those countries and find out what regulations/requirements there are for working in that country: visas, permits, etc.
3. **Find out** what companies are located in the countries you wish to work in, and whether they are solely based in that country or whether any of them have U.S. offices.
4. **Decide** on two or three companies that fit your interests and either find contact details for that company in the U.S. and enquire, or contact the national engineering body from point 2 and enquire about these organizations.

4. So Just What Are Employers Looking For?

> I have covered some of what young engineers want and aspire to. Now it's time to focus on the other side: what are employers looking for in their young engineers? It may surprise you to learn that, while important, your knowledge of Bernoulli's theorem and Quantum Dynamics may not be the greatest assets employers seek to gain from you. There's a whole raft of qualities, traits and attributes they are looking for; we will look at what they are in this chapter.

Starting with the basics --- be punctual. Just showing up to work on a consistent, regular basis is a good start. Moving on from there, such things as becoming proficient at what you've been hired for, having and displaying an interest in engineering, being a team player, and having good communication skills are some of the fundamentals. Leadership is an attribute always sought after, at the core of which is being pro-active and willing to accept responsibility. Finally, being a problem-solver, applying systems thinking, showing an appreciation for the business aspect, being able to manage your own time, and knowing where to concentrate your energies comprise the key criteria employers look for in new young engineers.

The basics

There is much written about the kind of skills that will be needed by the 21st century workforce. From *The New Commission on the Skills of the American Workforce*:

"Strong skills in English, mathematics, technology, and science, as well as literature, history, and the arts will be essential for many; beyond this, candidates will have to be comfortable with ideas and abstractions, good at both analysis and synthesis, creative and innovative, self-disciplined and well organized, able to learn very quickly and work well as a member of a team and have the flexibility to adapt quickly to frequent changes in the labor market as the shifts in the economy become ever faster and more dramatic."[1]

Here's another list[2] of attributes, this one from *The Engineer of 2020*:

- Strong analytical skills
- Practical ingenuity
- Creativity
- Good communication
- Mastered the principles of business and management
- Understand and practice the principles of leadership
- High ethical standards
- Strong sense of professionalism
- Dynamism, agility, resilience and flexibility
- Lifelong learners

My own list includes many of those shown above, but I'd like to go a bit further and explain why some of these skills are important.

Being good at what you're hired to do and satisfying the elements of your job statement are two things an employer looks for. In many cases (and I've been guilty of this) young engineers get so wrapped up in their own enthusiasm and desire to get involved in bigger and better things, they tend to look to tasks off in the distance but not those right at their feet. As the old saying goes, "You need to crawl before you can walk and walk before you can run." It's necessary to show an employer that you can satisfy the job statement you were hired for before you talk about becoming the CEO. By taking this approach, your superiors can see you dealing with tasks at your level and can gain confidence that you will be ready for the next challenge when the opportunity arises.

Interest in engineering

You may read this and think it's fairly obvious that if you are an engineer, you should be interested in engineering. However, you'd be surprised how often people enter engineering jobs without any real enthusiasm or interest in the subject. An employer will usually see this by the level of engagement you have in what you do, whether you seek to learn about your work outside of what is required, or if you do things without constantly having to be told. It's an oft-repeated statement that the best employees are those who have a passion for their subject matter. Insure that the work you are involved in is something you are passionate about, or you won't be working to your maximum potential or giving your employer 100 percent.

Team skills

You've no doubt heard this many times before: being able to work as a member of a team is important, and employers look for this ability in potential new hires. Our preparation as engineers doesn't always lend itself to working well with others. Engineering is typically a difficult course to gain admission to at university. While at high school, it demands devotion to the study of fundamental mathematical and scientific building blocks. While teachers, parents and the like can assist and guide you, you have to put in the work to succeed. Because of our formative environment, we engineers are conditioned to work on our own rather than in teams.

Engineers are also pre-disposed toward the use of the left-brain. This means we naturally avert from issues that are non-deterministic or outside the technical realm. Yet, take a look at the world around you: logic, right and wrong, or black and white tend not to be the way the world works in all its complexity. Therefore, in our initial years, we are often at a disadvantage when trying to apply a clinical, left-brained university approach to the fuzzy real world.

Let's look at it another way. Assume for a moment that you remain fiercely independent, refuse to work with others, and only do projects you can complete on your own. What would be the outcome? Well, the first thing such an approach does is distance you from the rest of your workplace. It's true, you would not be encumbered with imprecise non-technical issues, but you also would not have the benefit of a non-engineer's ideas and suggestions. You run the risk of becoming socially isolated at work. You may have brilliant ideas but limited opportunity to convey them to anyone, and your lack of participation in the broader work place may limit the extent to which people care about them anyway. In today's organizations, the majority of work and projects are done in a collaborative team environment where you have to work with others.

Communication

Communication is another critical skill for engineers. As Kalani Jones, an engineering vice president at Tachyon Inc., explains: "It's hard enough to find a good engineer; finding one who can lead a team and speak well in front of customers is really hard to find."[3]

The way engineers communicate with others is probably the biggest criticism other professionals have with us. If you cannot communicate effectively, you are rendered inert as an engineer, and your ideas will have little resonance with the people around you. Three common communication problems engineers have are:

1. the overuse of jargon,
2. the inability to be concise, and
3. an unwillingness or inability to explain complex issues in simple terms.

How to overcome these communication issues will be addressed later on.

Item 3 (an unwillingness or inability to explain complex issues in simple terms) has much to do with the mindset that "knowledge is power." Egos are often attached to positions and projects; engineer's chests swell with pride when they discuss a topic so difficult and convoluted that a non-engineer cannot possibly grasp it.

> **Communicating clearly**
>
> Jack Welch, the former CEO of GE, had a great way of cutting through cluttered communication. When addressing his senior managers, he would say: "Let's pretend we're in high school. Take me through the basics,"[4] and with that he'd expect an explanation using simple terms. This effectively tested his managers' understanding; if they couldn't do this, Welch concluded they didn't know what they were talking about.

An underlying cause behind such defensive posturing is the (wrongly-held) belief that if others can understand what's going on, our importance may be undermined, and our reason for being needed will disappear. Such thoughts are linked to the need for acceptance and status, and this is where the subject of personal and team leadership comes to the fore.

Leadership

How many times have you been on a project that seems to continue aimlessly with no coherent focus or direction? This usually occurs because no one steps up to take responsibility --- personal responsibility not only for their component of the project but for the project in general. One of the traits employers look for in an employee is a willingness to accept responsibility, and this isn't just for management positions. Leadership can be practiced throughout the organization, from the most junior staff member on up.

While many claim they want to be in charge, few have a good understanding of what taking a leadership position actually means. "Telling everyone else what to do" shows a limited understanding of what real leadership is. Giving orders without justification alienates people and, as I mentioned earlier, Gen X and Gen Y are not motivated by authority and intimidation.

The group meeting, a mainstay of most workplaces, tends to be a telling indicator of who accepts and who shirks responsibility. Deficiency in a chairperson's leadership skills is apparent when the meeting constantly veers off into areas that have nothing to do with the project, or when meetings run way over time. Both show a lack of concern for attendees --- yet, it's not all that difficult to prevent them from happening.

A second quality of leadership is pro-activity. Those who are pro-active automatically gravitate towards a leadership role because they tend to have thought things through ahead of time, or have more answers than anyone else. A solid agenda disseminated before a meeting, coupled with a suggested end time, is always a good start. I can't recall how many times at the conclusion of a meeting I would get this strange sensation that yet again nothing had been accomplished, no one was any wiser, and a large portion of my day was gone.

Cohesion and forward momentum in meetings is aided by all attendees being properly prepared --- but these principally result from a good set of action items at the end of the meeting, assigned to specific people, with defined time frames and expectations. Unfortunately, this detailing of required actions is typically where things fall apart, as actions often become so ambiguous as to be meaningless. "Joe is to get back to the group on Friday on the status of the E software module." An action item like that will usually elicit a last-minute response by Joe, as he's eating his lunch, such as: "In progress."

The third leadership trait is clarity. People need to understand exactly what they need to do, in no uncertain terms, if they are to have any chance of achieving the outcomes you desire. "Joe is to report back to the group via email no later than 12:00 Friday on 1) main areas of hold-up in E module code, 2) estimated date of completion for coding, 3) software testing schedule, 4) estimated prelim demonstration date." While the elements of what constitutes good reporting for code can be debated, in order to get precise details about how something is tracking you need to be precise about what you want, and in order to do that you need to know what it is you want. You can't expect people to read your mind and give you the answers you need; in all likelihood, the path of least resistance, or that required to "just get the job done," will be followed.

The final necessary leadership element is the setting of expectations. If expectations are established at the outset, contributors cannot use a lack of awareness as a recourse when they are not met. This is a common reason why projects run over time, over budget, and under requirements. Team members claim they "did not know what was expected." It's good to set your own expectations, even if your employer cannot or will not. This is important because it demonstrates your leadership skills in two ways: you're being pro-active by thinking through what your contribution is, and you've defined the height at which you've set the bar, giving your employer a standard against which to measure you.

Problem Solving

Every job, regardless of the position, will involve solving some kind of problem. Put another way, your employer is going to want you to provide a solution to a problem they have, be that a report they ask you to write, technical issue they ask you to explore, or project they ask you to lead. Therefore, problem solving is an important skill. Those with an aptitude for problem solving usually possess an innate curiosity: to find out how things work, to establish relationships of cause and effect, and to determine what the right questions to ask are. They don't start out with all of the answers but rather lay the foundation to ensure a process exists for getting them.

A true problem-solver should not be swayed by bias or second-hand uncorroborated information. They will seek to verify facts, usually from several different (and ideally independent) sources, ensuring consistency on multiple levels. Usually a spread of solutions comes from solving problems in such a manner, with the next step being to propose one or more solutions, all the while being mindful of things like your organization's resource constraints, political sensitivities, and trade-offs or alternatives.

Organizations are all about solving problems. If you relish tackling them instead of seeking to avoid them, you'll be highly valued in your organization. In fact, this is the best way to distinguish yourself from others; search out those issues that no one else wants to address and see it as a challenge.

Systems thinking

As innate curiosity is a characteristic of problem solvers, "systems thinking" is a good methodology for solving those problems.

"Systems thinking" is a way of seeing any complex situation as a system comprised of interconnections and relationships that can be broken down into simpler, more easily understood sub-systems. Such an approach to problem solving has been embraced by great thinkers from Leonardo da Vinci to the founder of modern management theory Dr. Edward Demming.

If we take a look at the rate at which technology has changed over the last few decades, one can start to appreciate why systems thinking is an increasingly important skill to possess.

Systems thinking and the telephone

Since its introduction in the late 1800s[5], the telephone has changed little in either form or function. Initially beginning its existence as a means for geographically distant (beyond line of site) parties to communicate, its design remained relatively unchanged until the 1980s when the first cellular mobile system came into operation[6]. Although the handset prototypes were large and suitcase-bound, they still provided the same function as their desktop brethren --- communication with those who were geographically separated. By the late 1980s and 1990s, a transformation in thinking emerged and the phone was no longer viewed solely as a means of transmitting voice. It was now viewed as a personal communication device, and companies scrambled (and are still scrambling) to arrive at new applications and modifications --- email, imagery, texting, and video to name but a few.

Perhaps you can begin to understand why systems thinking is so highly sought after. With each new application added to the cell phone, the new product extends the previous market, essentially becoming a "first mover" each time with no competition. As business environments like mobile telephony are being turned upside-down so quickly, industry participants no longer have the luxury of employing experts in particular areas to have them sit forever and a day dealing with discipline-specific problems. They now want employees who can be utilized in a number of different ways.

Being a systems thinker enables you to be an engineer in the truest sense of the word, to be innovative beyond just the technical domain. These people are useful not because of the information they have at any particular moment, but because of their ability to be flexible, adaptable and problem-focused, whether looking at the production processes in a canning plant, creating the project timeline for a new infrastructure development, or project managing a team to design a new integrated control system.

Systems thinking has a strong connection with systems engineering and is highly prized because of its connection to delivering projects on time, on budget and to specification. The President's commission on the Implementation of United States Exploration Policy[7], also known as The Aldridge Commission, sought to "examine and make recommendations"[8] on implementing the new vision of returning to the Moon and going on to Mars. One of the key findings from this report was that a "system-of-systems approach"[9] be used to assure NASA the best chance of meeting the President's goals within time, cost and specification constraints. Utilizing such an approach aligns closely with the System Engineering Mission statement jointly developed by The International Council on Systems Engineering (INCOSE) and the American Institute of Aeronautics and Astronautics System Engineering Technical Committee (AIAA SETC): to "assure the fully integrated development and realization of products which meet stakeholders' expectations within cost, schedule, and risk constraints."[10]

Business appreciation

As a newly hired engineer, it's natural to think that you'll be focusing primarily on technical issues and building your understanding in this area. Understanding the technology an organization is involved in is, however, only one piece of the puzzle. For example, to gain a true understanding of an organization's products and set yourself apart, it will be necessary for you to consider the environment that surrounds the company. Many factors will provide you with a much deeper, richer understanding and context for what you are working on. This includes your company's competitors, the industry you're in, the micro-economic environment at the time, and some of the macro-environmental events happening on the world stage.

Understanding the business aspects of IT

A lack of holistic understanding is a constant source of frustration for employers --- not just engineers, but for technical people in general. Referring to new IT professionals starting work for JP Morgan in Ireland, the head of the European Technology Centre commented: "My big frustration is that the graduates coming through our doors have no real understanding of the world of commerce."[11]

Time Management

The ability to manage competing requirements is an essential skill to have in the workplace. As our lives outside work involve more and more time-consuming activities, skilled time management has great utility. As an engineer in today's workforce, you will most likely either want or be expected to work on multiple tasks at the same time. Your level of involvement will likely increase with seniority, as your experience and wisdom are further leveraged.

Fortunately, most of today's young engineers are comfortable with simultaneous multiple tasks, as multi-tasking is part of their everyday lives. Seamlessly shifting between activities is more second nature than a skill they have to acquire, but just how desirable is such parallelism in a work environment?

In the *Stanford Daily*, a student commented on writing a 4500-word term paper: "I marvelled at how much longer it would have taken me to do this if I had been distracted by IM, email, Facebook, NYTimes.com, ESPN.com, Economist.com, friends, phone calls, rearranging my iTunes playlists – all frequent enablers of my procrastination."[12] While it is one thing to multi-task, your employer in the work environment needs an output of a certain standard within a certain timeframe. This means it is not enough to produce just *anything*; it has to meet a certain minimum standard. Regardless of how quickly you get through the task or how many units you have produced, if it doesn't meet that minimum standard it doesn't count as "partial credit."

The perils of multi-tasking

To provide an example of how output can be impacted by multi-tasking, let's look at the drop in performance that occurs when doing just two things at once. Repeated studies have shown that driving a car while talking on a cell phone massively reduces alertness and the ability to react, causing twice as many traffic signals to be missed, and taking longer to react to signals that are detected[13]. If doing just two things at the same time reduces the output below what is needed to stay alive in the context of driving, one can draw parallel conclusions in the workplace: unless actively managed, output will suffer as the task load increases.

Time management is about getting as much done in the available time rather than reaching a certain pre-defined level of time management mastery[14]. If you spend more time trying to optimize your time than you actually save as a result of the process, you are better off not having done it in the first place. What matters is results. Furthermore,

.you are better off getting one thing done well than a number of things done poorly. We want to meet an employer's needs by avoiding these two extremes: trying to do too much and failing to do anything satisfactorily, or focusing too much on any one thing and failing to attend to other tasks awaiting your attention. It's a fine balance that is dynamic and hard to keep in check.

The 80 percent rule

The 80 percent rule says that if a task only requires an 80 percent solution, then that's all you give it. One of the first things young engineers should do when given a task is ask the time frame given to complete it. This will usually be a good guide as to the level of detail that is required in the solution. As engineers, we typically tend to have trouble with this because our tendency is to provide the best possible technical solution as a way of reflecting the amount of creativity and innovation we possess. This is okay when we are at home and have all the time in the world to extract that last 1 percent of performance out of a motor, but rarely are time frames and resources unlimited in the workplace.

Any work you do will be based on a requirement accompanied by an expectation of quality. "Gold plating" a solution, typically considered a good thing by engineers, can be unwelcome for many customers. Take for example a product like sticky tape. What would be the point of providing 25 percent more adhesion than the closest name brand if yours and your competitors' product already met the customer's requirement? You could argue, "Who wouldn't want to provide the best quality product?" But such thinking dismisses who it is that determines this -- the customer. The young engineer needs to realize that when delivering a product or service to a customer, it's not *your* needs that are the focus, but *theirs*. Remembering this can save you a lot of time and the organization a lot of money, and will keep the customer happy because they've been listened to and their suggestions acted on.

Chapter 4, if you only have 5 minutes...

- There have been a number of reports (e.g., *The New Commission on the Skills of the American Workforce* and *The Engineer of 2020*) that detail the skills advantageous for engineers to possess in the 21st century.

- Being good at what you've been asked to do is a fundamental that young engineers often overlook in their attempts to prove what they can do.

- Having a passion for engineering is critical if you are to maximize your potential with any employer.

- Working well in teams is a necessary skill in the modern workplace, given that so much work is done within teams.

- Being an effective engineer requires effective communication that is concise, free from jargon, and reduces complex issues to their simplest form.

- Overly complex explanations (that few can understand) sometimes result from a deep need to feel important or valued.

- Leadership is less about position title and more about personal action and responsibility: the most junior employee can display leadership.

Actions to take away

Some things you can do to act on the information presented here:

1. **Take** the following actions to demonstrate the qualities, traits or attributes desired by employers:

Quality, trait or attribute	Action
Punctuality	Turn up to work on time on a consistent basis.
Being good at what you're hired to do	Ensure you satisfy the elements of your original job statement before branching out to seek higher duties.
Interest in engineering	If you're hired as an engineer and don't have a passion for it, it will show.
Team player	Work with other members of your immediate work group. Seek out and provide input to others.
Communication	Communicate technical ideas to your non-technical audience without the use of jargon, verbosity and unnecessary complexity.
Willingness to accept responsibility	Take ownership and display personal leadership whenever you can.
Pro-activity	Think things through ahead of time. Look at the situation from multiple perspectives.
Clarity	Clearly communicate so that your message is not only understood, but cannot be misunderstood.
Expectation setting	Let others know who you've assigned tasks to, and what your definition of success is.
Problem solving	Seek out objective answers, not just those that may be convenient or most easily found.
Systems thinking	Devolve a complex situation into its constituent components. Break up large, complex problems into smaller, simpler ones.
Business appreciation	Consider more than just technical issues, such as: competitors, micro and macro-economic environment
Time management	Manage your time well. Make sure that you deliver results.
Economy of effort	Know when it is worth putting in the extra effort, and when it will only yield marginal returns.

5. Job Hunting

> No matter what area of engineering you enter, or even if you don't go into engineering at all, sooner or later you're going to have to look for a job. There are a number of issues to consider when entering the job market. Understanding which industries align with your interests and what kinds of work are available within them is a good start. Next, an awareness of which industries are experiencing growth and what macro-economic events may lead to growth in the longer term will help you target environments that are most conducive to job seekers. The tools of the job search include the internet, friends, family, networks and professional societies. It is also important for job seekers to understand the role Human Resources and recruitment agencies play for their parent organizations. Finally, the important role of the resume and the interview conclude this section.

A changed environment

As previously mentioned, the new employer/employee relationship resulting from the turbulent 1980s and 1990s means you're likely to change your job, either within your organization or to another employer, every couple of years --- and if you're not, this could indicate that your skills are getting stale, or you're starting to settle down. Young engineers nowadays have a lot more freedom and opportunity to go after what they are really interested in. Along with this opportunity to flourish and succeed comes the potential to fall further and harder than at any previous time, without the comfort of an employer looking out for you.

So what's hot and what's not—where the jobs are

While identifying which sectors will grow and which won't is partly a product of how much you know about an industry, it is just as important to be aware of the trends in motion around the world impacting the industries you may wish to enter. Global Warming, the War on Terror, and Sustainability are all examples of global phenomena that have explicitly impacted many industries, increasing engineering employment opportunities within them.

Numerous articles in trade magazines and on the internet identify which sectors are hot and which are not. Remember, however, it is very difficult, particularly in the tech sector, to predict with any certainty what the long term trends are going to be --- things can change fast. An article from the Institute of Electrical and Electronics Engineers (IEEE) suggested the following areas would be seeing high levels of growth into the future[1]:

Biotechnology and bioengineering. Stem cell research and its application in curing diseases, such as Parkinson's and Alzheimer's, or to enable those with spinal cord injuries to walk, are going to be areas of great multi-disciplinary need. The development of drugs that more acutely target ailments is an area seeing many millions of dollars expended both in the R&D and commercialization phases.

Bio-technology

A specific example is the massive investment San Francisco saw based on the 2005 announcement that it would be the headquarters for the California Institute for Regenerative Medicine. There was a $3 billion dollar investment with a further $1.5 billion to be invested in upgrades to the University of California, San Francisco. The decision to base in San Francisco was the direct result of groundwork put in over the last two decades, making it a favorable investment opportunity with respect to taxation.[2] The potentially large rewards of having such a center has encouraged other cities around the world to try to attract emerging industries.

Engineers who understand not only the technology needed to create such drugs but how to then turn such products towards commercial profitability will be in great demand. Typically, those working in biotechnology with a background in research have little to no commercial experience, and therefore knowledge of issues like technology transfer can be extremely valuable. To meet this need, courses such as "Fundamentals of Technology Transfer," offered by the Praxis Organization in the U.K., are extremely valuable. "I am now far more aware of what industry might want from a license or collaboration," commented a Manchester university lecturer who teaches enterprise and advises university start ups.[3] Such courses provide tools for those trained in the technical disciplines to link more successfully with the customer, making them that much more valuable.

Alternative energy technology. Discovering new energy sources will require devising new means and methods of locomotion for cars, trucks, boats and trains. It will also require us to look at "undesirable" byproducts of modern living, such as excess rubbish or pollution, and find ways to put it to constructive use as alternative energy sources. Engineers employed in this area will need to think laterally in order to come up with new solutions.

New solutions do not necessarily mean new technology. Alternative fuels made from cornstarch and ethanol represent ways of using what we already have in our environment in a new way. A company in New Jersey, Terracycle, has taken the utilization of elements existing in our environment to a whole new level. They market a liquid plant food made primarily from worm waste by packaging it in old soda bottles, which helps keep costs down.[4] The characteristics of such a company speak to a central theme of this book: seeing the engineer as an innovator, not just the deliverer of new technology.

Many have asked why we're not putting more money into ethanol and using it as much as possible to replace fossil fuels. In a 2006 interview with CNN, the CEO of Chevron, the fifth largest oil company in the world, commented that while ethanol is meeting some component of demand, it is extracted from corn, and only so much corn can be grown given the limited amount of arable land. On the world stage at present, around 15 percent of the corn grown goes into a total of two percent of the fuel.[5]

Brazil

The use of ethanol to service certain markets is most evident in countries such as Brazil, where its use is booming. Brazil has been extremely progressive in embracing the use of ethanol; 10 years ago they invested $11 billion into its development, and the savings have been around $30 billion over the last 8 years. Unlike ethanol produced from corn in the U.S., Brazil produces its ethanol from sugar cane, which is up to four times more efficient.[6]

There are many other alternative energy sources that with increased development can provide useful energy yields. More money is being invested in these technologies as they edge closer and closer to commercial viability: solar, biofuels, wind and biomass are a few of the more common examples being developed.[7] While naysayers are quick to point out that alternative energy sources are impractical, necessity (being the mother of invention) will force us to find more economically palatable alternatives to fossil fuels. Investing in such technologies provides organizations an opportunity to consider their environmental impact, minimize their environmental footprint (or, at least, the public's perception of it) while at the same time pursuing cost reductions. This means being environmentally conscious can, in fact, have a positive impact on the bottom line.

In his 2006 State of the Union address, President Bush decried America's "Addiction to Oil." Many saw this as an indication that, after years of limping along, alternative energy's time had finally come. With Hurricane Katrina sending fuel prices skyrocketing in the U.S., the environment was perfect for hybrid cars to enter the market, and enter they did. Languishing in the shadow of conventional fossil fuels, alternative energy technology has always lacked conviction due to a shortage of political will. This has now changed.

Consumers have signalled that automotive giants like GM need to recognize their demand for cars that are fuel efficient and highly reliable. Although GM was successful in winning back some customers with "employee pricing," (customers pay what GM employees pay), the "Buy American" loyalty they counted on for years was not absolute – it held only as long as the quality of American-made cars was equal to other cars on the market, and that simply isn't the case.

Most of the major car manufacturers either have current hybrid models or models on the drawing board. In a symbolic gesture, Toyota released a hybrid version of the Camry, one of world's most popular cars. And while better gas mileage, safety records and tax breaks provide incentives for consumers to make the switch, the economic situation isn't right at the moment. Hybrids comprised only 1.5% of sales in 2006. And although it is projected hybrid sales will make up 5% of automotive sales by 2013, this is well short of what was expected earlier on. Nissan CEO Carlos Ghosn commented: "Consumers are finding hybrids don't save enough gas to justify the extra expense."[8] The increased market demand for non-fossil fuel alternatives can only be met by drawing in more engineers to work on every stage of the life cycle for new car designs. There have already been some new and innovative concepts, such as the Automated Vehicle Health Management system (Onstar), in which the driver receives regular emails from the car notifying them of the current performance of various systems such as vehicle emissions, ABS brakes, tire pressure, mileage and so on.[9] If a crisis of global proportion is to be averted, we must lessen our reliance on fossil fuels --- not just for transportation, but in all aspects of its use in our economy.

Biometrics and security. The Global War on Terror was born with the terrorist attacks on the World Trade Center on 9/11. A switch from passive acceptance to a more pro-active stance by a host of nations (led primarily by the U.S.) has highlighted a desire by many to stay ahead of terrorist networks. This has led to a large increase in the resources available for developments utilized in the War on Terror. Therefore, engineers who wish to be involved in high technology will find a growth industry in defense. Of all sectors, defense is the most likely to grow for the foreseeable future, at least until there is a major shift in the global geopolitical situation. The biggest drawback for engineers wishing to work in the defense sector is the required clearances, typically proof of citizenship and other background checks. Therefore, despite the need to source the best and brightest, defense and homeland security jobs are likely to be restricted not only to citizens, but also to those who can obtain relevant security clearances.

Civil and infrastructure. As our cities become increasingly crowded and expand their boundaries, more high rise buildings will be constructed. Increased population will further exacerbate demands on water supplies and other infrastructure that support these mega-cities. Nations around the world are finally realizing their civil infrastructure, allowed to decay for too long, is in dire need of repair and upgrade. Billions of dollars need to be poured into these infrastructures, and all of this work will require engineers --- not only in the development and construction phases, but in the maintenance of support systems, and support of legacy buildings and systems that will be around for years to come as their greatest possible utility is extracted before being upgraded.

U.S. bridges in need of repair

In August 2007, 13 people died and 145 people were injured when the I-35W bridge spanning the Mississippi River collapsed[10]. Design error and the weight of construction materials used during resurfacing were responsible[11].

As tragic as the I-35 bridge collapse was, more disquieting is the report, *Bridging the Gap: Restoring and Rebuilding the Nations Bridges,* released by the American Association of State and Highway Transportation Officials. The report states: "Almost one in four bridges, while safe to travel, is either structurally deficient, in need of repair, or functionally obsolete, which means they are too narrow for today's traffic volumes."[12] There are 590,000 bridges in the United States, and it would take $140 billion in funding to fix them all. Clearly, with major investment being considered, a workforce element including jobs for engineers will be needed.

Internet Services. From the tangibly large scale to the intangibly large scale, the development of the internet will continue to march forward and the latest wave to sweep over it is the provision of web services, commonly known as Cloud computing. This means previously "shrink wrapped" applications, like Microsoft Word, will be transitioned into an online service. This is a sound business model for companies like Microsoft to exploit; instead of charging a one-time fee for buying an application, users would be charged a monthly fee. From the users' perspective, web services represent a change for the better as they eliminate the need to upgrade to new versions of software --- they are upgraded on the server end and the user, if paid up to date, transitions over to the upgraded product. Delivering such web services will require engineers at all levels, not only to develop the software but to incorporate feedback from users on what they want and how to make the service better. Looking ahead, web service organizations will also be trying to lead consumer demand, with the further exploitation of mobile devices currently being the hottest area for web services development.

There are many other sectors that will experience growth in the coming years, but the important point here is that the engineer as innovator will have application in literally any problem solving environment. Therefore, future employment prospects for engineers are very bright.

I've identified a sector, now what?

So you've weighed your interests with your undergraduate degree and identified the industry you'd like to work in. Now it's time to find out what jobs are out there. No matter which area of engineering you go into, one thing is key: to succeed in it long term you've got to have passion. You've got to really like what it is you're doing, and to see the challenge in it --- otherwise it's going to feel like work.

A good place to start looking for a job is through friends or those already working in the industry. Most people, however, start with the internet.

When you first look at online job postings and read what employers are looking for, you may think: "What planet do these superbly trained and experienced people come from?" The descriptions always seem to ask for a wide range of varied experiences, implying you need to have had three different careers running in parallel to qualify for their job.

If you feel completely confused by all the resume advice out there, don't worry. When I started my job search, my resume was in complete disarray. It took me an eternity before I began to realize why my resume was never being looked at, and I was never able to get anyone to tell me openly and honestly why I was never getting any responses. For a start, a resume and a curriculum vitae (CV) are not the same thing. (Don't worry, I never knew the difference between the two until it was explained to me by a colleague.)

A CV is essentially an account of everything you've done in your professional life. It is typically about 5-7 pages in length, and not aimed specificially to any one position or application. Contrast this with a resume that is usually 1-2 pages at most, constructed with a specific position or application in mind, and does not contain a complete account of all of your professional experience. In a resume, you take your experience and include those items relevant to the job you're applying for, and leave out those that have no bearing on it.

Most European countries and Australia tend to look at CVs because there'll usually only be a few applicants for a position, giving the reviewer time to go through them in detail. In the U.S. and other countries, where there may be hundreds or even thousands (yes, thousands) of applicants for one job, a resume is the only way to go. The reviewer spends a minimal amount of time with each resume and typically uses it as a way to cull out those who will not make the first cut. In the U.S., a resume will not win you the job --- it is intended to get you into a face to face interview where you can then tell them exactly why you're the best person for the job.

Human Resources

Obtaining a job with any organization means that, at some stage, you'll need to deal with the Human Resources department, or HR. Unfortunately, HR has a bad reputation with those looking for jobs, as well as with those within a company looking for the right people. From both viewpoints, regulation and adherence to policy are seen to obstruct the desired end result of hiring the right person for the job.

HR *should* be joined at the hip with the highest levels of the organization because they are the means, the primary unit responsible, for providing the human capital through which the organization satisfies its strategic objectives. Yet it can often appear as though HR does not operate with the company's strategic objectives uppermost in their mind when staffing the organization.

Perception of HR's disconnectedness

Thornton A. May, a management consultant and columnist for *Computerworld* magazine, lamented in a 2006 article: "HR in many organizations seems both disconnected from the goals of the corporation and insensitive, almost to the point of malfeasance, to the emerging needs of a highly skilled IT workforce."[13]

Once upon a time, companies hired people using their own recruiting staff. They had plenty of trained internal employees, so when they advertised a position they were able to sort through the responses, consider them with due diligence, and respond. Those days are long gone and unlikely to return so many organizations have neither the time or inclination to have their own HR staff do the initial screening. So, they outsource it in the belief that this will create value by receiving a short list of high quality, high calibre candidates. Unfortunately, this does not always make for a happy ending.

Outsourcing the sourcing of candidates is the last thing a company should do if they are serious about getting good quality talent. While it might seem logical to hand this part of the hiring process to specialists whose sole job is to find candidates for employers, in most cases recruiters do not understand the engineering environment and the different skill sets required. The frustration generated when dealing with many of these recruiters often drives applicants to try and find ways around them and go straight to the source. However, the employer has hired the recruiter to not only select the most appropriate candidates --- they have also been hired to keep applicants at arm's length. Even if you attempt an end run around the recruiter, the company, upon receipt of your resume, will likely send it back to them --- unless you know someone on the inside of the company who can usher it through to the inner sanctum[14].

My dealings with recruiters have not been overwhelmingly positive. One particular case illustrates how I believe the right candidate for the right job can be found if some time and effort is taken up front. I had applied for a number of positions at a particular company but never received a response from HR; emails and phone calls went unanswered. I had a contact within the company who had walked my resume in front of the hiring manager, but still nothing. "Recruiting has now been outsourced," my contact said, and I followed his advice to approach the outsourcing company. Unfortunately, because the outsourcing company was simply acting on what was written in the job description, they could not offer any advice.

After several months of applying and getting no response, I left it alone for a month or two, during which time I secured a permanent position through a contact I had at another company. Sometime later, I received a phone call from the first company informing me there was a position I was being considered for. I mentioned I had obtained a permanent position with another company, but that I would be happy to provide feedback on my dealings with them. "Please do" was the response. Over the course of a 30-minute telephone conversation, I told him that because I could not find out further information on what they had been looking for (other than the extreme technical requirements in the job description, which were poor at best) it made it impossible to apply for the position unless one had performed that exact job somewhere else (see previous section on "just in time" jobs). He commented that he had received a similar response from another person they interviewed at the company. They said that the job they went for and the job described were completely separate things.

This highlights an all-too-common problem created when job descriptions are not clearly expressed --- what a company would "like to have" versus what they "need to have" results in recruiters deflecting candidates who would otherwise be a good fit. This happens because despite the rhetoric of wanting to find the right people, organizations may not know what they're looking for.

Most recruiters operate on a commission basis, paid by the organization when a candidate is hired. As such, they are working for the company (not the candidate), and that is where their allegiance lies. Recruiting companies are primarily concerned in placing a person into a job --- not necessarily the best person. They may not be motivated to distinguish between candidates, or go the extra mile to get the best candidate; they are satisfied with one that merely meets the requirements. The best candidates and the ones that end up going to interview may not be one and the same.

The other alternative, the career management agency, is not necessarily any better. My experience with such an agency left me having to go to court to try and get my money back. So be careful: if you decide to utilize a career management agency, be aware there are many suspect ones out there. However, most states in the U.S. have public career management organizations whose services you can utilize for free as part of work relocation or work assistance programs.

Internal HR – fundamentally broken

The only way an organization can accomplish anything is through its people. This view is standard among organizations who at least want to be seen as forward looking. Without people, an organization is just an abstraction without substance. People are the only tools an organization has to make things happen. Therefore, you'd think that if you wanted to make an organization great, you'd stock it with the best kind of people.

This is where practice and theory part ways. Organizations may not always hire and retain the right kind of people because (1) they are overwhelmed by everyday business, and (2) they do not address the real issues that cause good employees to slip through the organization and out the other side (see previous section on "Toxic companies").

First, let's start with Human Resources, commonly known as HR. If you apply for a job as an engineer with any sizeable company, you're going to be faced with a numbers game, and the trick is to create a situation where you can whittle the numbers down to just one applicant: you. This doesn't tend to happen when you apply for jobs over the internet, and it requires you to make connections on the inside. You need to cultivate the idea that you're the only person who really understands the problems they're facing, and better yet, that you're the only one who has the tools necessary to provide solutions.

Globalization of the workforce has become a double-edged sword for organizations looking for employees, and for employees considering who they wish to work for. This expansion has caused HR departments to receive overwhelming responses to job vacancies. To put some figures on this: the global labor market went from 960 million in the 1980s to around 3 billion in 2000. While part of this is due to an increase in world population, a significant contributor has been the addition of economies such as China, India and the former Soviet Union to the world stage. Therefore, while you can now look to work anywhere in the world, the rest of the world can also look to your corner of the world for jobs. So, given you have access to a vastly greater network of jobs and job applicants, how are you going to access them quickly, easily and effectively? The answer is the internet.

The Internet

The use of the internet both for job seekers and job posters has meant a far greater field of view being available than ever before. Job seekers are now far more empowered by being able to apply for multiple jobs at the push of a button. But, how many responses have you received from employers when you blasted your resume to 50 companies using monster.com? Not many, I would imagine.

For job posters, the internet was seen as a great way to automate the search for talent. Simply ask potential employees to enter their details through your secure company web page, taking special note to highlight their particular job-relevant skills, and whenever a job comes up matching their requirements an automated email message is sent to them.

Great idea, if not for the inundation of resumes for each position, and the online fatigue for job seekers who have to enter their details into a new online form every time they want to apply for a job with a different company. (Sometimes these forms run into pages and pages, and even when you try and stay one step ahead by having your resume formatted to cut and paste into the relevant fields, you get timed out and have to start the process all over again.) Companies get thousands if not hundreds of thousands of resumes from people they are not interested in, and they end up frustrating those who they potentially would want. While the internet provides access to potential talent, it also significantly raises the noise floor. Not a very customer-focused process.

Most companies use automated search algorithms to sift through the resumes contained in their database, linking up keywords and phrases. Your task is no longer one of applying for a job, but rather one of being a "resume engineer," passing through all the right filters and setting off as many keyword alarms as possible in the hope you'll get noticed. You can now start to appreciate how it may not be the best candidate who gets the job, but the one whose resume is engineered towards hitting the job filters' buttons. There's got to be a better option for success, and in my opinion that option is knowing someone on the inside.

As engineers, it is naturally assumed we are accomplished at using the internet. Hence, it is not surprising to see that a host of online tools have sprung up allowing us to network in cyberspace: Plaxo, Linkedin and Ryze to name but a few. These allow you to connect with others based on profiles you can view. Akin to a dating service for professionals, they allow you to display testimonials from co-workers and colleagues, a powerful feature in particular of Linkedin. Instead of collecting references from everyone you've previously worked with and for, for every job you apply for, you can now ask them to submit it once, and attach it to your profile for any prospective employer to see. This has led to "reciprocal back slapping," where you ask one of your references to put in a testimonial and you, usually through a sense of obligation, write one for them which in turn is glowing. It becomes a testimonial arms race, with people collecting testimonials that declare they are the next Bill Gates or Donald Trump. When you have 8 or 10 of these super-amazing testimonials, it makes prospective employers a little suspect. Sometimes less is more.

Despite this, online networking tools have flourished; some employers look exclusively at candidates with testimonials from co-workers and colleagues. This bypasses those who get the manufactured reference or put in the last minute call to a former boss to "put in a good word" because a prospective employer is going to be calling on them.

The Insider

Having someone on the inside of a company can be worth its weight in gold when applying for a position.[15] They can not only give you the inside scoop on how applications are really processed (which keywords in your resume will set off the alarms) but in addition, they can often hand-deliver your resume to the hiring manager and give it an extra personal endorsement. "Hey Joe (or Susan), I know this guy (or gal) and I think they'd be perfect for job X" --- that's usually all it takes to lift you to the top of what can be a very tall and insurmountable pile if you're applying solely online. It's like trying to climb the pyramids from step one versus having a helicopter take you straight to the top.

If you don't know someone on the inside, the best way to meet one is through a networking event. There's usually no shortage of these events and conferences sponsored by a multitude of sources such as professional societies, chambers of commerce and industry groups. Sometimes there may be a small fee attached to cover the costs of catering. They often feature a guest speaker to talk about something relevant to that group. In meeting people at the event, you'll be able to start the conversation by reflecting on some of the key issues raised by the speaker.

The Interview

Unfortunately, interviewing often seems like the last thing on companies' minds when they conduct interviews for engineers. Perhaps the everyday workplace demands on people's time is to blame, but in most interviews I've had for engineering positions the interviewer was either not prepared or did not understand what it was they were looking for to fill the position.

This experience is a massive turn off for potential employees. If a company wants to get the best applicant for a position they should make a good first impression. While organizations may *say* they want the best candidate for a job, they often represent themselves in the interview process to the contrary. This is a major reason why many companies, despite screaming out for the "right kind of people," are coming up empty handed. The "right kind of people" are being turned off by what appears to be the "wrong type of company." In fact, studies have shown that "high performance organizations employ rigorous selection procedures that have been refined and developed over time to identify people with the attitudes and skills they require."[16] In other words, the right companies do their homework and know what it is they are looking for.

A colleague of mine interviewed with a large software company in 2005. Over a period of several weeks he went through something like 10 interviews. It finally became clear they weren't interviewing him to see if he was the correct fit, rather they were looking for some reason to justify why they weren't going to hire him. In continually bringing on the wrong type of people because of poor interview techniques, organizations have become so risk averse to hiring that they would rather carry a vacancy than fill it with someone who meets 80 percent of the criteria. Bewildering behavior.

This risk averse behavior in the recruiting process stems in part from the fear of potential litigation. Organizations are so conscious of being sued, and of the negative impact such an action could have on their public image, that many will do whatever they can to avoid that situation. Better to annoy a few job seekers than have a public court case and a tarnished image broadcast over the internet.

The purpose of an interview (for a permanent position) is for the organization to find out if what they see on paper matches the person they see in front of them. They need to find out not only if this person can do the job, but whether or not they can be an asset to the organization. Issues such as their longevity, and the value they might add to the company beyond their initial appointment, are important considerations.

A good interview should be a mixture of *specific job-related* and *generic attitudinal* questions. Many commentators on HR issues tend to lean towards specific job-related questions as the best predictors of success on the job, but I disagree. Attitude, a willingness to learn, and the fit with the organization should win out in today's environment. Don't get me wrong --- if the organization wants an engineer, then that person should have an engineering degree and some specific skills to match; but in many cases, organizations focus too much on technical skill sets when they really need holistic employees.

In my interviews, I always made it a policy to know not just about the organization, but also to show them I could connect the dots. When I interviewed with a large aerospace company here in the U.S., I discussed how the company was going to deliver continued profitability in the future, what the product mix was transitioning to, and why and how I thought the macro-economic climate was going to move over the coming years. Now you might suggest this was a bit much for a graduate fresh out of university to deliver convincingly, and that might be true, but if you at least give them a glimpse of your thinking outside of the typical technical lines most engineers will give them, then you'll at least set yourself apart from other applicants.

Finally, asking questions is key. This is something you should make a point to think about ahead of time. It is extremely difficult to come up with questions on the spot when asked. Asking the interviewer how your career path will unfold, what main areas they are working on, and what the biggest challenges they are facing all show them you are actually interested in the position, the organization and the work you are going to do for them.

Sometimes, despite all of this, the interview is not a main determinant in getting the job. Organizations may just want to see who's out there, or they may already have a candidate in mind. Being aware of this can take the pressure off, making your responses less guarded and scripted, which at least will give you a more natural demeanor in what can otherwise be a stressful experience for engineers in their early years.

5—Job Hunting

Chapter 5, if you only have 5 minutes...

- Today's workforce is characterized by more frequent job role changes.
- Greater freedom of job choice brings with it a greater opportunity to fail, given you no longer have a long term employer behind you.
- Identifying and understanding global trends (the Environment, War on Terror and Sustainability) can highlight which industries may be impacted and hence may be experiencing growth.
- The application of engineering towards the relief and cure of human ailments like Parkinson's disease and overcoming spinal cord injury means bio-technology and bio-engineering will be growth areas.
- The drive to discover new forms of energy, emergence of China and India from developing to developed countries, and the economic turmoil of fossil fuel prices all lead to the continued and sustained growth in the alternative energy technology area.
- The Global War on Terror has led to an increased focus on the provision of solutions in this area. Industries such as Defense, Aerospace, Biometrics and Security are but a few of the many beneficiaries of this focus.
- The need to continue the provision of basic sanitation, clean water and the maintenance and upkeep of the nation's civil infrastructure will be a growth area for many years.
- The growth in web services will require engineer's involvement at every stage of the lifecycle and will see the transition of previously "shrink-wrapped" products to the online domain.
- Friends, family, the internet, professional societies, conferences and networking events all represent viable mediums to explore for potential jobs.
- Resumes should be written for a particular job in mind and be 1 to 2 pages in length
- Human Resources should be aligned with the highest levels within any organization, as they are the means to obtaining the people through which the organization gets things done.
- Outsourcing of candidate selection to recruitment firms has increased the frustration of many job seekers.
- Job descriptions as written rarely reflect what a job role actually entails or what the successful applicant will actually do.
- Recruiters hired by organizations have an allegiance to the

- company to place a suitable candidate. This should not be confused with placing a specific job seeker into a job.
- Trying to take an end run around the recruiting company and going directly to the organization may end with you back dealing with the recruiter.
- Organizations may not acquire and retain the best talent through a combination of being overwhelmed by more immediate business and not addressing the real reasons behind why people leave.
- Developing connections and networks is a good way to tap into the real job market, not just the one advertised on the internet.
- Workforce globalization has increased the global labor pool from 960 million in 1980s to 3 billion in 2000.
- Sending the same resume to 50 different companies through monster.com or similar sites is unlikely to yield a high response rate.
- Applying through online portals for jobs shifts your role from job seeker to resume engineer, since your resume usually needs to be selected based on keywords in order to get through.
- Online network sites such as LinkedIn or Plaxo are good additions to face to face networking, but shouldn't be used as the primary means.
- Organizations that put forward unprepared interviewers do themselves a dis-service as it reflects poorly on them.
- The litigious environment means that organizations are ultra-wary of showing favoritism or bias in the hiring process
- The resume gets you into the interview, the interview gets you the job.
- Asking questions in an interview shows you have an interest in the organization you're hoping to work for.
- Some organizations may not be looking to fill a position, rather just see what applicants are out there.

Actions to take away

Some things you can do to act on the information presented here:

Job Searching

1. **Write** down all of those things that you are interested in. Do this for 60 seconds.
2. **Consider** which industries out there appeal to you. Take another 60 seconds.
3. **Match** the elements that you are interested with in your first list to the industries that you listed in your second. What can you say about this matching? Do all of your interests align with one particular industry, or are they spread across multiple ones?
4. **Approach** a friend, colleague or family member over the coming week and ask if they know anyone associated with the area(s) you are interested in. If they do, ask for an introduction with the view to discussing your interest(s) in that field and ultimately finding a job within it.

Networking

1. **Find** a professional society that is aligned with your interest or engineering discipline. If you do not know of any, look at the list provided in the references section.
2. **Send** an email to either a local chapter representative or point of contact listed on the website, and ask them about networking events and seminars in your area.
3. **Go** to one event within two weeks of finding a professional society or group you are interested in. Make sure you listen to what is being said and try to take away at least one thing from the event.
4. **Engage** with people after the event. Ask them:
    a. What field of engineering they are involved with.
    b. What they like/find challenging about what they do.
    c. How they got involved with that field.
    d. If it sounds like an area that is of interest to you, ask them for a business card.

Engineering Challenges

1. **Go** to http://www.engineeringchallenges.org and watch the "The Grand Challenges for Engineering" video.
2. **Ask** yourself whether you agree with the challenges put forth by those in the video. Are there others that you can think of that are not mentioned?

6. Being An Engineer—So What's It Like?

> In this chapter, we'll cover the concept of being an engineer versus becoming one. Being an engineer is more about process than result. Engineers are not a homogeneous group --- they fall into a number of different categories. In this chapter we'll consider four of them: the technical purist, the Project Manager, the academic, and the entrepreneur. Each category brings with it certain advantages and disadvantages which must be considered when deciding which category to identify with.

Being versus Becoming

While I was at university, becoming an engineer seemed very distant. You graduated university with an engineering degree, and suddenly you were an engineer! After I completed my degree and entered the workforce (in my case the Air Force) I realized that I didn't all of a sudden become an engineer. In an odd way, the title came first --- and the "becoming" was something that continues to this day, and will probably continue throughout my whole professional life.

This process of "becoming" an engineer started when I began looking at the world in a certain way, influenced by the education I had received. It reminds me of the movie "The Matrix." Throughout the film, Morpheus leads Neo towards seeing the world as it really is --- an artificial construct, a sea of binary numbers zipping around him. Becoming an engineer is somewhat like that for me: when walking around the world, I now tend to see things from the engineering perspective: whether it be viewing a bridge and wondering how the design keeps it from falling over, looking at a telecommunications tower and imagining it radiating radio waves that blanket the sky, or envisaging new product combinations that reduce costs of production and satisfy customers demands in new and innovative ways.

Embracing the non-technical

This engineering perspective includes more than just the technical aspects. As previously discussed, engineering is a rich, multi-disciplined endeavor, not one merely wrapped up in specific, narrow disciplines. Many people have the pre-conceived notion that as an engineer you're a techie and therefore sit at your computer in your own world designing things that may or may not have relevance to the real world. More and more, however, engineers are required to not only be technically proficient but also have an understanding of other disciplines such as finance, law, economics, workplace relations, etc.

Unfortunately, many engineers have a poor view of others in the profession who seek to expand on their non-technical skills. They view with disdain the pollution of their technical prowess with the ungodliness of finance or law. Notice that I said the engineer must have an *understanding* of other disciplines --- this doesn't mean you have to be able to mix it up with the best legal eagles or Wall Street analysts. You should, however, be able to understand your legal obligations, or know how to read a balance sheet. While you obviously want to retain your technical core, jobs like manager in a technically-based organization will require you to keep your engineering wits about you while being ever mindful of important non-technical issues.

The power industry serves as a good example of the engineer's need to understand disciplines outside engineering. This industry currently faces issues involving the regulatory environment. A comment in the feedback section of IEEE's *The Institute* magazine complained: "Those who are now making the regulatory and investment decisions don't even try to understand the technical consequences of what they do. We should invest prudently in power and communications systems and rely on qualified power engineers empowered to make the appropriate decisions."[1] Many engineers in the industry are frustrated by what they see as a lack of awareness of new regulations and their impact on operations. Understanding this, and seeing the potential such an opportunity brings, a number of engineers have made quite a lot of money consulting to those who set regulations, providing an insider's perspective on the operational impacts of such changes as well as deciphering technical jargon into laymen's terms.

While the media and Hollywood may portray engineers as stereotypical geeks with poor social skills and love of technology for its own sake, there are in fact almost an infinite number of jobs that engineers perform. This makes it hard for people to understand what engineers do; they cover a vast number of industries, and two people at the same point in their careers in the same industry can be working in very different jobs. There is no "typical engineer," but there are a few categories that engineers can be placed into.

The technical purist

When I was going through university trying to make sense of the onslaught of differential equations, quantum theory, Kirchoff's current law and the myriad of other rules, laws and equations, I gained a new-found respect for those who called themselves "engineers." I would ask myself, "How on earth did they ever survive this?" Then I came to the even worse realization: how am I ever going to survive a whole career in this profession if I'm fighting this hard to get through my undergraduate degree? While some quickly step over the detailed technical elements after graduation, others make it their life work.

It is often said that the engineering subjects you study throughout your degree, in all their depth and complexity, are those you'll be confronted with throughout your career. This view makes some salivate with excitement, while others shriek in terror.

An engineer who holds fast to the technical purist role can take one of two paths: the general or the specific purist. Those who seek to be technical generalists want only to deal with technical issues, just not necessarily the same ones all of the time. They may, for example, know how to cut a bit of code, understand some circuit design, and know about simulation.

This kind of position can be filled with a lot of dynamism: rotating around different tasks, dealing with differing concepts. However, this mode of existence has a lifespan of only a few years. Organizations rotate new engineers through different positions on short-focused projects to give them a taste of what they have to offer, expecting the employee to pick an area of interest or focus soon after the initial period of rotations. This brings us to the second avenue: being a niche technical purist.

The knowledge of a niche technical purist is necessarily going to be limited to a defined field, but they will know the ins and outs of everything to do with that subject. They will probably be known as a Subject Matter Expert, and are the type of people you meet at technical seminars who are the go-to guys (or gals) with respect to resistors, particular bridge spans, or the one who is the Encyclopaedia Britannica on control systems for unmanned underwater vehicles. These people are usually more than a few years into their careers, due to the fact that it takes time to build up expertise in a particular area. They probably, though not always, have a higher degree, possibly a PhD, and they may also be a Technical Fellow at the organization they work for.

So, what are the pros and cons of being a technical purist, someone who only wants to deal with the technical issues? The upside is that you get to be revered as somewhat of a demigod in your particular field of expertise. You'll probably expand the knowledge base in your area, get to work from sun-up to sun-down on a specific area --- particularly if you're acting in some kind of research and development capacity --- and you'll probably interact with many of the world's authorities in that area. There is, however, a downside.

The downside of being such a specialist in either a general or niche sense is that you may feel a little removed from those in your organization --- distant from those who are not so technically literate, but also disconnected from the organization's direction and lacking any ability to change it. Also, you may not have the chance to engage with the end customer. If you're working on a product for market, a customer engineer may act as a buffer between you and the end customer, or at least those who are providing the requirements.

Another negative: you may find your greatest strength (your expertise) turns into your greatest weakness as you specialize yourself out of a job. If the market shifts and the organization discontinues the area you're an expert in, you may find you've worked in one niche area for so long --- and developed no other skills that the employer can utilize – that you are no longer any use to your employer.

Finally, your role in the organization may be marginalized by co-workers and management. Staying a purely technical person may earn you the label of "geek or "techie," often seen as someone unable to contribute to the direction of the organization. As a technical purist, it is possible the direction and intent of the organization may not even interest you, but highly technical engineers often feel sidelined by management. There's a good reason for this, however: engineering is there to support the business need, not drive it. At this point in your career you may hit a glass ceiling that will require you to develop other skill sets to penetrate, perhaps heading you in the direction of becoming a Project Manager. But beware, not all engineers make good Project Managers.

The Project Manager

A Project Manager is something of a special creature. Years of service alone does not guarantee success, although experience does help. Few engineers become truly effective Project Managers, because going to this next level is not merely "more of the same" but requires a different mode of thinking.

This different mode of thinking can be illustrated by results from the Myers-Briggs personality test. This battery of questions provides a four-letter code representing a specific personality type from sixteen possible combinations. The typical engineer is classified INTJ: Introverted, Intuitive, Thinking, and Judging. The typical manager, however, is classified ENTJ: Extroverted, Intuitive, Thinking, Judging.

Note the difference here between "E" (Extroverted) and "I" (Introverted). Instead of sitting in the background and processing thoughts quietly, the engineer must now externalize them, think aloud and be very verbal. This is something many engineers abhor because we tend not to open our mouths until we have something sensible or decisive to say. Anything less than this and it seems like we're shooting from the hip.

> **Engineers to managers**
>
> The traits of an INTJ are developed and strengthened from the very moment one enters engineering at university, particularly through those courses of study such as computer and electrical engineering. Professor Ray Findlay, an Emeritus Professor of Electrical and Computer Engineering, has taught undergraduates at McMaster University in Canada since 1981. Observing over many years who successfully makes the transition to management, he's found a strong correlation with Myers-Briggs[2]; "It's a fact that not that many engineers can make the transition into management."[3]

So just what is a Project Manager? A Project Manager is a facilitator, someone who pulls a technical team together and keeps their eye on the bigger picture of cost, schedule and performance. They are in the middle of a larger web, with responsibilities both up and down, left and right. They have a responsibility to report the status of the project to those above them, and to raise issues from the team that cannot be dealt with or resolved at the Project Manager's level. At the same time, they take information from upper levels and feed it to the team in an appropriate and meaningful form. They communicate left and right with different projects that may be impacted by, and who may impact, their project. They may also interface outside the organization to an ultimate end user or customer.

I think of Project Managers as executing "bottom-up management," where they are at the apex of an upside-down pyramid. They are there to provide guidance when needed, but at the same time lay down a path for the team to follow. The Project Manager needs to have enough of a grasp of the information to understand what's going on, but at the same time needs to sit on top of the information.

They need to be aware of who within their organization possesses the knowledge, and be the catalyst for tapping into that when the need arises.

Given the Project Manager's central position in the hub, you may have guessed their greatest asset is communication, both in written and verbal form. But it's not just about being able to speak up and concisely convey a viewpoint to a wide variety of audiences. It is also just as important for the Project Manager to be a good listener, one who listens with empathy. The very definition of management is getting things done through other people; therefore if the Project Manager is not a good listener, it is unlikely those on the team will be willing to carry out his or her wishes, at least not to the best of their ability. The importance of listening, and knowing when to let another speak without interruption, cannot be overstated.

Unfortunately, listening skills tend to be lacking in the workplace. They are not only useful for Project Managers, but for all job types. As Project Managers, many engineers forget their new role and give in to their primal urge for solving detailed technical problems. They are so used to dealing in technical issues that instead of helping others to solve them, which engenders a sense of confidence and self-reliance in the team, they yank it away from them and dive into the intricate details as though they were back as part of the technical team. This usually results in team members feeling sidelined, redundant and frustrated. The Project Manager is now doing two job descriptions instead of one, and probably neither very well.

Micro-management is another problem that can occur with Project Managers who have an engineering background. There can be nothing more annoying, frustrating and downright demoralizing than working under such a person, having them re-do your work or monitor every step you take.

Micro-management

I experienced micro-management first hand as an engineer in the Air Force, and ended up losing many nights sleep over it. Over a period of weeks I handed work in to my Commanding Officer, and he'd go ahead and re-do it without telling me why. After pondering it for a while, I finally gathered up the courage to broach the subject with him. He was genuinely surprised to hear that his actions were being perceived as "micro-management." He said he just wanted the work to be of high quality. A fair enough sentiment, but unfortunately for some people no one else's work is ever good enough.

While many purist engineers view the Project Manager as a non-engineering role, I would argue to the contrary. Instead of optimizing for a technical solution, the Project Manager has a broader requirement to optimize a solution over multiple domains, with a core focus on cost, time and performance. When viewed in this light, one could argue it is more an engineering role than the technical engineer, particularly in today's world where an appreciation at a systems level is so important.

Greg Walters, a manager with the global professional services company Sinclair Knight Merz, suggests the position of Project Manager depends largely on what role the incumbent is to perform. "Too often engineers call themselves Project Managers when they should more correctly call themselves Engineering Managers."[4] He suggests that the greater the complexity and size of the project to be dealt with, the earlier in their career the engineer needs to start acquiring project management skills, and increasingly the client management and softer skills as well. "The bigger the project, the more critical those skills become," says Walters, who has worked on many projects for both public and private sector clients[5].

Because of the huge impact they have on the outcome of a project, Project Managers are in high demand and will continue to be so for the foreseeable future. For this reason, companies should never seek to outsource the Project Manager role. While a shortage of technical talent is hurting organizations, Project Management skill sets are also hard to find. "Project Managers will never be commoditized," comments Ross Dawson, Chairman of the Future Exploration Network, an organization devoted to the development of strategies and information capabilities to provide competitive advantage[6].

The Academic

Some engineers decide to spend their career in the university environment. They are excited by the possibility of conducting research, diving into specific engineering areas and charting new territory. However, the road to becoming an engineering academic is not an easy one.

Take the example of Dr. Richard Radke, an assistant electrical, computer and systems engineering professor at Renssalaer Polytechnic Institute in New York. As he was nearing the end of his PhD, a senior professor he knew took him aside and told him: "I hate to tell you this, but it's going to be brutal."[7] Those aren't the kind of words you want to hear when you're starting down the path towards full tenure.

One must jump through many hoops to get tenure at universities in the United States which, some say, shifts the focus from producing quality research to producing papers in order to "publish or perish." Recognizing this perception, universities in the U.K., Japan, India, China and some other Asian countries have abolished tenure and given academics fixed term appointments. This frees them from the need to secure their positions, and allows them to focus on their research.

The imperative to publish speaks to another problem most undergraduates are familiar with. With their professor's survival linked to research, teaching comes in a distant second. "If you're an excellent researcher and a so-so teacher, you're okay. If you're a so-so researcher and an excellent teacher that's not going to fly," comments Russ Joseph, assistant professor in electrical engineering and computer science at Northwestern University[8].

Academics are also criticized for failing to engender a sense of entrepreneurialism in tomorrow's engineers. This may be a problem that takes some time to address, for those engineers inexperienced in industry do not possess the background to pass this on to undergraduates. Also, for those academics who have had experience spinning off companies from the university sector based on research they have done, there may be the problem that such spin-offs are necessarily technology centric --- e.g., development of a new modem or router, as opposed to something of a more generic nature so students can be more holistic in their entrepreneurial approach to engineering. They need to appreciate that technology needs to satisfy the business need and not be developed for its own sake.

The Entrepreneur

At first glance, the methods of many entrepreneurs may appear crass to engineers. Usually their thinking can be tangential, discontinuous and replete with abrupt changes in direction, quite annoying to an engineer used to structure and a logical path. But if you remember earlier on, we derived the word engineer from its Latin root, *ingenium,* which means *to innovate.* This means that engineers creativity and lateralism should be sitting right alongside that other part of them that takes ideas and systematically figures out how they can be turned into reality.

Unfortunately, there are very few examples of engineers who successfully make the transition to entrepreneur because of the change in focus it requires. The engineer deals predominantly with issues technical in nature, whereas the entrepreneur, similar to the Project Manager, while needing to be cogniscent of the technical has to have a purview not only broader in scope but extending along another axis; that is, the lifecycle of whatever product or service they are trying to sell[9].

This is one of the reasons why entrepreneurial engineers typically don't spend much time in large companies; they tend to feel stifled and hemmed in. Unless they are part of a smaller group, they often feel they don't have the freedom to be innovative and creative. Their passion and drive makes them be appear single-minded. This can hinder their advancement in large organizations, where time spent within the enterprise becomes a greater concern than in smaller organizations.

In his book, *Lead Like an Entrepreneur: Keeping the Entrepreneurial Spirit Alive Within the Corporation,* Neal Thornberry, Professor of Management at Babson College in Massachusetts, describes entrepreneurs not so much in terms of what they do, but more by the attitude they possess. "I think it's a certain mindset. Add to that an emotion that focuses on discovery to try and find something new, and if you see it, to build a business out of it."[10]

Being entrepreneurial isn't necessarily about having the latest and best whiz-bang gadget; this is where most engineers depart the scene. Many engineers seem far more wedded to the technology rather than the problem the technology is trying to solve. They spend their time coming up with technically superior products while considering the customer's needs secondary. The entrepreneur, however, views the product or service as a medium through which they satisfy the customer's need. A creative solution is developed by looking at what exists now, what may have existed in the past, and how they can be brought together in innovative and creative ways to satisfy a present need. That's a display of entrepreneurial spirit[11].

Take for example the Nike + iPod sport kit. Nike developed a sensor you place in your shoe that connects wirelessly to your iPod Nano. Runners use the iPod as a data logger and tracker for parameters such as distance, time and speed, all via the sensor placed inside Nike shoes[12]. Your iPod keeps track of the calories you burn, as well as provides motivational messages that increase in intensity as you near a pre-designated finish. This is a perfect example of how bringing together pre-existing products can satisfy a need without developing something from scratch.

While the entrepreneurial spirit seems alive and well in the U.S., there is a big push abroad to get entrepreneurship to be seen as something to aspire to, and that can solve a number of society's problems. Unfortunately, old habits die hard. A recent survey of French students found that "only a tiny number want to go the entrepreneurial route, while 75 percent fancy life-long civil servitude."[13] As our public institutions struggle under the burden of increased demands and reduced budgets, it is imperative an entrepreneurial mindset be allowed to flourish and grow. It is only with this kind of attitude that continual creativity can take place, and new and better ways to solve problems discovered. Engineers have a large role to play in developing such solutions.

Chapter 6, if you only have 5 minutes...

- Becoming an engineer is a process that starts with graduation and extends throughout your career.
- The inclusion of finance, law and other non-technical skills should not be seen as taking away from the technical skill set, rather an enhancement to it.
- There is no typical day in the life of an engineer; two people at the same point in their careers in the same industry can be working two very different jobs.
- While engineers work in a myriad of different roles and job titles, some categories they fit into are: technical purist, Project Manager, academic and entrepreneur.
- The technical purist is an engineer who specializes in either a specific or broad technical area.
- **Advantages**:
    - Highly revered and renowned in the field.
    - Considered a subject matter expert.
    - Constantly work on your area of expertise.
- **Disadvantages:**
    - Possible feeling of disconnectedness with organization's direction.
    - Chance that skill set becomes irrelevant to employer.
- The Project Manager is one who leads a project, satisfying a customer need within cost, schedule and performance constraints.
- **Advantages:**
    - Connected with many different levels of the organization.
    - Calls on very broad based cross-section of skills to get the job done.

- **Disadvantages**:
    - Feelings of frustration through a possible lack of ownership of resources.
    - Can be a high pressure position with a high degree of accountability and responsibility.
- The academic is an engineer who works in an academic, research, and/or teaching position.
- **Advantages**:
    - Continue delving into those areas you first touched on in your undergraduate degree.
    - Once tenure is reached, job security with lifelong employment.
- **Disadvantages**:
    - Very tough road to tenure.
    - Pressure to produce research papers.
    - Usually requires many more years of study, masters and PhD
- The entrepreneur is an engineer who seeks to discover new means and methods of satisfying various stakeholders.
- **Advantages**:
    - Be your own boss.
    - Can be very lucrative financially.
    - Not stifled by bureaucracy.
- **Disadvantages**:
    - No safety or employer looking out for you.
    - Usually high pressure
    - Usually not a logical or coherent process from start to end with many frustrating dead-end paths along the way.

Actions to take away

Some things you can do to act on the information presented here:

Which category of engineer?

1. **Take** 60 seconds and write down all of the things you would like to do as an engineer.
2. **Re-read** the four broad categories listed in the chapter: technical purist, Project Manager, academic and entrepreneur.
3. **Write** down which category each point or element of your list falls into. You may find that they span more than one category.
4. **Ask** yourself what this means? For example, what does it mean to have elements within both the entrepreneurial and project management categories?
5. **Think** about the type of jobs that may utilize these cross category elements.

Skill sets

1. **Find** someone from a professional society or organization who you think fits into one or other of these categories. For example, you could seek out a past lecturer or someone who is a Technical Society Chair.
2. **Query** them on how they see their category. Do they see themselves fitting neatly under the umbrella title of Academic or Project Manager? If not, why not?
3. **Pose** questions to this contact on their skill set. What non-technical skills do they use and how do they view accruing skills that reside outside the traditional technical areas?
4. **Think** about the statement "Becoming an engineer is a process that starts with graduation and extends throughout your career." Why do you think this statement is valid or not?

The View From Here

7. Communication—The Engineer's Lynchpin

> Communication is a fundamental skill required from all roles and at all levels of an organization. In this chapter, we'll address the components of communication: verbal, written and visual, and suggest some steps we can take to make improvements. Powerpoint has become a popular presentation tool, yet like any tool it can be abused. We'll look at how it can be best utilized so your presentations are clear and on message. Written communication is more than just putting pen to paper; it's a process involving a number of steps that, when followed, result in a written piece free of superfluous language, spelling mistakes or logical inconsistencies. I'll offer up a process that will help you achieve written works that will stand out. Finally, your communication style shifts depending on your audience --- whether it be the general public, your colleagues or a senior manager. Being mindful of this can help you shape your message in the most effective way.

Good Communication—So what?

Everyone says that good communication is important, but why? Communication is the only way we can transmit our ideas and thoughts to others. What and how we communicate form other peoples' opinions of us. If we look closely at communication, we'll find it comprises the verbal, written and visual elements; therefore, in order to be better communicators we need to consider all of its sub-components.

Our verbal communication has a powerful impact on people. What we say and the way we say it determines whether people stop and listen or continue talking and take no notice. Your ability to speak well is not something that automatically accrues with seniority. I have attended many a conference where the head of a professional organization spoke in such a way as to be completely and instantly forgettable, and yet I've also listened to speakers who, holding no high position of office, were engaging and commanding. It's something that needs to be practiced, and the most likely forum for us as engineers to do this in is the presentation.

Most people hate giving presentations and because they hate it they shy away from it, in turn exacerbating the problem to the point where they become deathly afraid of it. However, giving a good presentation is nothing magical. It comes from being prepared, understanding your subject matter, and thorough rehearsal (unless, of course, it's a completely impromptu presentation).

I not only wanted to become good at giving presentations, I was also willing to find out what it was like to be bad at them. I wanted to put myself in harm's way and see how people reacted to the different things I tried. As I have always had a fascination for, and some level of understanding about, the space industry, I thought it was only natural that I use this subject matter as a platform to test out my skills (or lack thereof) and present a series of space seminars to anyone who would listen. I presented to such forums as engineering students at the Australian National University (ANU), the Rotary Club, the Aviation Club, Questacon, ITEE, the Royal Aeronautical Society and Engineers Australia. At each presentation, I handed out feedback sheets and asked the audience to put down as many of their likes and dislikes as possible, no matter how small, picky or insignificant. After the presentation, I would collect them and compile a two or three page report for myself. I invited a colleague to run a fine tooth comb over my delivery and tell me what he thought I needed work on, no matter how insignificant. Doing this required me to be completely open and non-defensive, as some of the comments I received were harsh, unreasonable and downright unfair. However, with each pursuant presentation I tried to remember what I had learned from the previous one and consciously alter the disliked behavior. At the very least, doing these presentations gave me exposure as a speaker and practice at delivering a message to an audience. While I am always striving for improvement, I am now much more aware of my tendencies and can give a presentation without getting too stressed out.

We've all been to some absolutely shocking presentations, especially early on in our careers at university. Some of the more common mistakes include: not making eye contact with the audience, mumbling or not speaking clearly, talking too slow or too fast, and speaking to the screen or blackboard. There seems to be a perception that public speaking comes naturally, but it is a skill like any other which can only be developed by practice, and practice does make perfect.

A good way of practicing is to deliver your presentation to yourself in the mirror. While this may sound silly, I can attest that it works. It is a powerful method of practice in that you can make eye contact with yourself and see first hand what you do when you speak --- i.e., play with your hair, shift your weight from one foot to another, fiddle with your hands, etc. Once you can deliver a talk to yourself in the mirror without fidgeting or reading from your notes, you'll be ready to give that speech to any audience.

Powerpoint – evil to its core or useful tool?

If you're like most people in engineering and science, you are no doubt familiar with Microsoft Powerpoint, the universal presentation tool. Despite its widespread use, there are those who detest it. Among these detractors is Professor Edward Tuft, a Yale political scientist and specialist in the visual display of information. In his 2003 book, *The Cognitive Style of Powerpoint,* he argued that the use of Powerpoint encourages "statistical stupidity" and "turns everything into a sales pitch"[1]. A reasonable enough stance, but not a harsh enough indictment according to other critics such as Ruth Marcus, a journalist for *The Washington Post.* She went one step further, suggesting Powerpoint was more dangerous than putting us to sleep at boring presentations. The opening line of her column boldly asks: "Did Powerpoint make the space shuttle crash?,"[2] hinting at the kind of blame game that tends to follow Powerpoint.

If we stand back and look at this rationally, Powerpoint is just a tool we use to communicate. It can't conspire, it can't give orders, and it can't tell you to cut the red wire over the blue one. Before Powerpoint, we had slide shows and clear acetates onto which a confusing mass of hand-drawn charts and diagrams were placed. Powerpoint is not responsible for convoluted presentations --- we've been doing that with our charts and diagrams long before its invention. At least Powerpoint's fonts and styles give us a sense of consistency throughout a presentation. Powerpoint is not the problem --- the real problems are found further upstream.

Time and again detractors of Powerpoint (primarily scientists and engineers) suggest that using the tool constrains the speaker and doesn't allow them to get into the proper flow of presenting data. As an engineer, I understand the need for hard facts and quantifiable data in order to support certain points of view. However, many scientists and engineers use Powerpoint more as a speaking crutch rather than a speaking aid. As Professor Tuft commented: "PowerPoint is a competent slide manager and projector. But rather than supplementing a presentation, it has become a substitute for it."[3]. Enough said; Powerpoint is a means for getting your point across, but don't let it dictate what your point is.

Passive presentations are those whereby nothing is demanded

of the speaker or audience. We've all done it. Gone along to a presentation you know little or nothing about, or not prepared for a meeting by skipping the background material. We all hope and pray that the speaker will be our saving grace, that their presentation will contain all the information we need, and that it'll be easy to follow. Unfortunately, this is rarely the case and letting the speaker get away with this says to those in the room, "I don't value your time enough to have prepared what I'm going to say and you don't value yours enough to call me on it". Although we are all starved for time in our busy worlds, it is astounding the amount of people who give Powerpoint presentations without doing the most basic preparation. You can always tell: they read straight from the slides, go way over time, do not engage the audience, and basically leave you to make your own conclusions on the data they have presented.

Therefore, for presentations to work properly, both sides have to engage. You need an audience who's looked at the background material beforehand and brings some issues they want addressed to the table, and on the other side you need a speaker who knows his or her stuff and who's going to use Powerpoint to highlight their main points, not as their teleprompter.

Such an approach takes work --- it takes an active response from both parties to say: "I'm going to be prepared." Unfortunately, due to time constraints, most of us succumb to last minute winging it or the back of the room snooze. As there are so few who prepare for these events, you'll likely get bonus points at work if you do prepare, whether you're a speaker or a listener.

Writing

Ask engineers why they went into engineering and a common response will be: "Because I don't like writing." Upon entering engineering, most students think they can breathe a sigh of relief and forget about ever having to write again. Unfortunately, this is not the case. Your ability to communicate in the written form will be called upon on a daily basis, more so than your technical knowledge. Probably the most profound class I took during my years of study was called Engineering Practice. It consisted of engineering graduates telling us about their workplace experiences. One woman stood up and told us the single most important skill we would use and be constantly judged on would be writing. As I went through my initial years in the workplace, this turned out to be very true.

Why is this the case? Aren't engineers concerned with solving technical problems, and doesn't writing and documentation take them away from this? It's true, you will solve your fair share of technical problems; however, there are very few jobs as an engineer where you work in isolation. You almost always have to communicate your results to someone, be it in a formal journal, a paper for a conference, or to a less technical audience --- perhaps in a meeting to your boss or some other senior managers.

You might say, "Well so what? Who cares if I don't write well? As an engineer, I'll be judged on my technical prowess, and once they see what a star I am at that my writing skills won't matter." Wrong, wrong and wrong again. Even if you work in complete isolation, there's always going to be times you'll need to come back to work you completed earlier or refresh yourself with what you've previously done. If all you've got are a few scrawls on a bit of paper, you'll be hard pressed to take on long term projects because you'll never be able to keep it all in your head. At some stage you'll have to write it all down, and that's when you'll be in trouble. Clear written communication is a must.

There are many guides out there on good writing, and I'll leave it to you to find them. In my opinion, good writing comes from first having an understanding of what it is you are trying to say, and then crafting it. I write in an iterative fashion. If I have an article or assignment to write, I don't sit down at the computer at one in the morning the day its due, write it out start to finish, get the spell checker to correct mistakes, then hand it in. Instead I take a layered approach (an approach, by the way, arrived at by use of the one in the morning all nighter).

First, if it's an assignment, I'll put down the word limit and a range of ten percent. This means if you have an essay of 2000 words, a range from 1800 to 2200 words will be acceptable. Second, as most people use a standard 12 point font and single line spacing, I work out how many pages this is, with a rough rule of thumb of 500 words to a page. This is probably an underestimation, but you can work out an easy estimate which will make your document end up being anywhere from 3.5 to just over 4 pages in length for a 2000 word piece. Next, factor in the introduction and conclusion, each 250 to 300 words, and you're left with around 1200 to 1600 words to address the subject of your assignment or report.

After mapping the structure, next comes writing the document. I usually write a series of main points or areas that I may want to talk about, then under each of these a few dot points. If I'm really stuck for a starting point, say when I'm writing an open article on a topic, I'll just put down all of the things related to it on a page and then from there move towards assembling them under common headings. I'll then start to look at how each of these subject areas link together, looking at how I can flow one point to the next.

In writing, you have to take your audience on a journey. You show them the case you are presenting, then lead them down a path of facts or issues such that your conclusion is the only logical one that could be reached based on what you've provided. There are no surprises and no loose ends.

To do this, you'll have to do multiple drafts of your document. OHHHH!!!!! I hear you say, Do I have to??!!!! Well yes, you do, and here's a few reasons why. You need to write a number of drafts because to get to a polished end product you have to link up sections, re-word arguments, and strike out superfluous language. When writing your document the first time, you don't want to obstruct the flow of your ideas by worrying about spelling, grammar, or even if it remotely makes sense to anyone else. You won't be able to correct spelling, link up sections and re-word arguments all at the same time, so you have to take each in turn, going through the document several times, checking for different things each time. It's important to note that because spell-checkers cannot ensure correct context, they cannot ensure correct spelling. The amount of times I've seen the word "your" used in place of "you're" is beyond recollection. This only has to happen once and your document has the stamp of an illiterate. If you take the time to draft and go through it thoughtfully, you'll have a flowing, logical and coherent document that's been carefully constructed, is a work of quality and considered forethought, and uses just the required amount of words to explain your ideas.

> **Casuistry**
>
> Blaise Pascal, the 17th century mathematician and physicist most famous for developing the concept of pressure, attacked a commonly used method of reasoning for his time known as *casuistry*. Casuistry was basically verbose or excessive reasoning used with the intent to mislead. What Pascal was attacking was the reasoning methods of various members of the Catholic Church. He saw that such methods did nothing to inform or educate, obscuring the need for true reasoning on the writer's behalf and replacing it with long-winded arguments designed to hide a lack of knowledge. In his most famously quoted assault on casuistry, Pascal wrote: "I would have written a shorter letter, but I did not have the time."[4] Pascal's implication here was that it takes longer to craft a short letter than it does to write a long one.

Plagiarism

The issue of plagiarism is related to our earlier discussion about cheating in engineering. While you've likely heard of plagiarism, you may not know exactly what it is. My education did not provide a concise definition of what plagiarism is and what it is not.

While "copying someone else's work" is the most widespread view of plagiarism, it in fact extends beyond that to include copying someone else's *ideas* and not acknowledging them as the original source.

Plagiarism in industry

A high profile case of plagiarism occurred in the U.S. in 2006. Raytheon CEO William Swanson wrote a book called *Swanson's Unwritten Rules of Management*, but soon after it was written it was found to contain a large number of similarities to another book titled, *The Unwritten Laws of Engineering,* penned in 1944 by a UCLA engineering professor. While Swanson apologized for the indiscretion, it cost him a raise and stock that he otherwise would have received from the Raytheon board of directors.[5]

There is nothing wrong with using other people's ideas. In fact, a famous phrase you've probably heard illustrates the legitimacy of leveraging off of the discoveries of those who have come before us: "If I have seen further than others, it is only because I have stood on the shoulders of giants."[6] These words uttered by Sir Isaac Newton referred to the fact that his discovery of the universal law of gravitation depended heavily on the groundbreaking work done by his predecessor Johannes Kepler.

Leveraging is not a bad thing in itself, as without it society would never progress; we would find it necessary to continually re-invent the wheel. The automobile, for example, has gone through hundreds if not thousands of incremental improvements, each one building on the past to give us the cars we drive today. But the lesson here for engineers is that it is not ethical to utilize the ideas of others and not credit or cite them for it.

Visual

The way we look and act when first meeting a person influences to a large extent how they connect with us. Upon meeting someone new, your opinion of them is usually fashioned by the way they look and act, not necessarily by what they say. Although there are only a discrete number of words you can get out upon first meeting someone, appearance and image are instantaneous, and a whole raft of opinions are formed concerning their social status, how much money they make, whether they're sophisticated or uncultured, and so on. Being mindful of this fact, especially when meeting someone for the first time, is important.

How we look

Scientists and engineers are often stereotyped in the popular media as being goofy, socially inept, or awkward. Just look at examples such as the "Nutty Professor" (either the original with Jerry Lewis or the remake with Eddie Murphy), the "Dilbert" cartoon series, and any of the "Back to the Future" movies.

While there are a lot of inaccuracies in the way engineers are portrayed, there is contained within a kernel of truth. You only have to walk around on a lunch break in Silicon Valley to see very quickly who works in software or has a technically focused job. Sandals, white socks, loud stripey tops and tracksuit pants are all too common, distinguishing features. This might irk you a bit --- since when does what we wear determine our usefulness to the organization we work for? That's more the domain of marketing or those PR types, isn't it?

While many engineers have a resistance to focusing on what they wear, and some may not care how others perceive them, perception is everything and even if you don't care about it, someone else will and it will influence their opinion of you, like it or not. Dress such as that mentioned above creates an image of being somewhat of a misfit, therefore it should come as no surprise to engineers who dress in this or another unprofessional fashion when they are passed up for promotion, representative positions, or jobs that interface with the customer.

While the rise of software companies in the dot com era brought a swing towards the casual workplace, dress is starting to swing back the other way. Organizations are re-introducing policies of "smart casual." While I'm not advocating that everyone wear a suit, some kind of pride in your appearance does outwardly portray a level of professionalism, suggesting to the observer that because you have an attention to detail in how you dress, you're more likely to have attention to detail in your work as well. Therefore, taking time to put away that cartoon tie or leave the Simpson's socks in the drawer can have a measurable positive result.

What we do

How you dress will obviously depend on the kind of work you do. For operational roles, such as a mining engineer at a mine site constantly inspecting crushers, interfacing with technical tradesmen on site, and entering and exiting control rooms, it's probably not a good idea to wear your best Armani suit. Usually there'll be a particular uniform, dress or standard that everyone adheres to, putting the issue of wardrobe out of your hands.

The environment you work in will also influence your job description, in turn dictating how you dress. In an office environment, you will likely interface with customers or clients, using your skills of persuasion to get your point of view across. In an operationally focused position, you may be more task focused, concentrating on specific details needed to get the job done. The kind of people you interface with to get your message across will also have a large role to play in how you communicate, and that's what we'll concentrate on next.

Audiences

Senior managers often complain that engineers needlessly ramble on and on about small level details, so insignificant and irrelevant as to have no impact on any bigger level issue. They constantly stress the need for strategic thought and seeing the "bigger picture."

A colleague of mine who was the CEO of a large not-for-profit organization told me about an example from his past that illustrates this point. One of the engineers in his organization wanted to speak to him about an issue of great importance. He agreed and had the engineer come to his office. He soon found himself becoming increasingly annoyed as the engineer droned on and on about some small technical detail rife with acronyms and pseudo-speak. Before long, my colleague tuned out the engineer and started checking his email. The engineer kept going. Finally, he interrupted the engineer and told him that what he was talking about was really a problem to be dealt with at his level, was of no concern at the upper levels, and while he had found it infinitely fascinating, unless he had bigger issues to discuss the conversation was over. The engineer, it seems, could not understand why others did not find what he was talking about as important as he did.

This may be a problem with engineers who are knee deep in the technical details. When speaking with others, it is always necessary to be mindful of who you are talking to, and to modify the message accordingly.

To your peers

Your peers are those at the same professional level and standing in the organization as you. In most cases, these are the people you can discuss intricate technical detail with until you are blue in the face, because they work on your level. Their knowledge will overlap yours in certain areas, but in others you will know about certain things they don't, and vice versa. There is usually a large degree of familiarity, so to an outsider much of what you discuss would seem laden with acronyms and insider talk.

To your immediate boss

The industry you work in will determine to a large extent the degree to which your boss is knowledgeable about what you are working on. For example, in the software industry, a senior programmer will probably be expected to know all that a junior programmer knows and more. The junior programmer's knowledge will be a subset of the knowledge of the senior programmer, and not the other way around. That may not be the case in other industries where the junior engineer may have more specific technical knowledge, but not a greater level of overall, general knowledge.

Usually, the explicit tasks you are given to complete will be assigned by your immediate boss. How good a boss they are and their comfort level with your mastery of the material you deal with will determine the amount of micro-management you experience. A bad boss, or one who has no confidence in your abilities, will dictate how you are to do something, and use how he or she would do it as a determinant in whether you've done an acceptable job. Therefore, when engineers rise in seniority and have others working for them, they need to let go of the details and allow others' imaginative input to effectively contribute to how tasks are done. So long as the goals are achieved, personal creativity should be allowed.

A good boss will tell you they need something done --- how you do it is up to you. This allows you to bring your creative side to bear, giving you a great deal of satisfaction and personal input. There are cases, however, when it is necessary to dictate how something is to be done; for example, in cases where a certain process must be followed as dictated by the customer, or for certain dangerous or hazardous processes where deviating from a checklist could cause serious harm.

Communication from you to your boss should be to the point, relevant, and done with a solution in mind. They are looking for people who can come to them with solutions. Problems are not bad in themselves, but you should have a range of possible solutions ready to offer. Despite what you may think, your boss does want to know what you are working on; unless they're working alongside you, you are likely the best judge of the situation and of what the next steps should be. If you come across a problem, let them know why it is a sticking point. If you don't, their ability to provide guidance or make a decision is hampered by their lack of information. Too many times people hide problems or setbacks because they fear reprisal or failure, but the only way we learn is to make mistakes. Remember, that person who is your boss today was sitting where you are not that long ago.

To a senior manager

The higher up an organization you go, the more strategic your outlook becomes and the more encompassing your field of concern is. Just think of climbing a mountain; if you were at the top, you'd be able to see for miles in every direction but probably not make out too many of the details. If you were at the bottom of the mountain, you'd see things like individual trees, rocks and maybe even local fauna, but you couldn't know the view from above. So it is when engaging senior managers. A Program Manager's or Vice President's time is necessarily going to be limited because they are responsible for a number of projects, perhaps even one or two major programs. Also, their focus is going to be different than yours. Program Managers are going to be focused on how the overall program is tracking with regards to cost, schedule and performance. If issues you have are going to impact any of these three at the program level (not the project level) then they'll want to hear about it. They won't want to hear the detailed technical argument, although they will probably want a synopsis of that. They'll want you to get straight to the point and connect the dots as to why the impact is there and what your opinion of mitigating solutions may be.

To the public

When speaking to the general public, it's a good idea not to assume the audience has a high level of knowledge concerning what you're talking about. This is not to say you should treat people like morons --- just try to explain things fully, in clear language they can understand, and with passion. Nothing turns people off more than hearing a dry technical explanation on a topic they know little about. Lots of diagrams and illustrations help explain concepts and get everyone on the same page. If people feel like they're genuinely learning about a topic that is made interesting to them or which they can see has application in some way, they'll be a receptive audience.

Chapter 7, if you only have 5 minutes...

- Communicating is the only way in which ideas can be transmitted to others, hence the reason why good communication is so important.
- Communication is comprised of written, verbal and visual elements.
- Good communication does not accrue with seniority, it is a skill that must be learned and practiced.
- Common mistakes in presentations include: lack of eye contact, mumbling or not speaking clearly, talking too slow or too fast, and not addressing the audience.
- A method of practicing for presentations is to rehearse in front of a mirror and take note of your mannerisms. Make sure to look yourself in the eye.
- Powerpoint should be seen as an aid to delivering your presentation, not an excuse for why it cannot be more effective.
- When delivering presentations, show respect for your audience by being prepared and sticking to your time limit.
- No matter how technically brilliant you are, good written communication is necessary to transmit ideas to others as well as provide a record so you do not need to store everything in your head.
- Good writing requires a layered approach, which implies multiple drafts before arriving at the final product.
- Some elements of writing an article or assignment are scoping the word limit, factoring in an introduction and conclusion, getting down unstructured and nebulous thoughts on paper and then connecting them in a logical sequence.
- Spell-checkers are a useful tool, but they should not replace proofreading your work for out of context words, such as "you're" instead of "your."

- Plagiarism is not only copying someone else's work without crediting them, but also their ideas.
- A person's appearance transmits perceptions to others far more than any other forms of communication do.
- Paying attention to your presentation in the workplace sends an implicit signal to your employer that you'll pay similar attention to your work.
- A person's vocation dictates what is appropriate dress.
- Understanding your audience is necessary to tailoring the content of your message.
- Communicating with peers is likely to be full of acronyms, pseudo-speak and insider language.
- Communicating with your immediate boss should be relevant, to the point and always done with a solution in mind.
- Your boss really does want your input given that their actions may be largely influenced by it.
- Communicating to a senior manager should be done with their perspective in mind. They most likely will take in a much broader perspective than you are aware of.
- Communicating with the general public should be simple but not simplistic. The use of diagrams and illustrations will help to clearly articulate your point.

Actions to take away

Some things you can do to act on the information presented here:

Verbal communication

1. **Critique** your own verbal communication. How would you say you speak to people? Do you tend to take too much time when delivering a message, or do you rush through to get it all out?

2. **List** some areas that you think need improvement in your verbal communication. Some examples may be: speaking up, slowing down, or leaving pauses between sentences to allow other people to speak.

3. **Record** yourself giving a presentation on whatever topic you like. Either use a webcam, the record function on a digital camera, or a video camera. If you do not have access to any of these instruments, give your presentation in front of a mirror.

4. **Watch** the recording and write down all of the things you see that need improvement or that could be considered distracting. How does this list compare with the one you made earlier? Are they the same or different?

5. **Ask** someone else to watch the recording (best friend, parent, colleague) and to put down or tell you their thoughts on what they think needs improving. How does their list compare with the one you generated?

Written communication

1. **Look** for any assignments or written work that have been marked in the past on which comments regarding your written communications have been placed. Compare them and see if they are commenting on the same or different things. Have you made active attempts to address this, or does the same area keep coming up?

2. **Try** the following steps for your next assignment or written work:
   a. Put down the word limit +/- 10%; this is your range.
   b. Divide the range by 500 to get a rough estimate of how many pages that equates to when typed on the computer.
   c. Subtract 600 from the word limit to account for the introduction and conclusion (each 300 words).
   d. Consider the question being asked. Do you have to analyse, interpret, compare, evaluate or undertake some other action?
   e. For 3 minutes, brainstorm all the elements you can think of that may be related to answering the question.
   f. Assemble the elements into like groups; that is, ones that relate to one another.
   g. Try to put these groups in some kind of logical order (if possible).
   h. Start fleshing out some of the main points under each of the groups with actual sentences. Keep building until you've done this for all of your points.
   i. Type what you have so far into your computer. Don't worry about looking for spelling mistakes or any other corrections at this stage.
   j. Once typed, print out a copy (double-sided if possible to save on paper).
   k. Read what you have printed out. Does what you've written make sense? Do the ideas actually connect, and is there superfluous language in there?
   l. With red-pen in hand, write in any corrections you have in terms of argument structure, circle spelling mistakes and correct grammar.
   m. Once the body is shaped up, write your conclusion. The conclusion should not introduce any new material but provide a summary of your main arguments and results.

n. Once your conclusion is complete, focus on your introduction. The introduction shouldn't go in depth about anything in the body of your work but lightly touch on every argument or issue you are going to raise.

8. Outsourcing And Off-Shoring

> There are many issues and trends sweeping across the globe that alter the way we work, interact with others and move forward. While you will have undoubtedly seen some of these issues in the media, and may have formed opinions of your own, you may not have considered them as a whole. It is important to understand or at least be aware of phenomena such as outsourcing, off-shoring, global mobility, value-adding services, the utility of politics in the workplace, increased importance of collaboration, ethics and diversity. While they are outside of the technical realm, they are a feature of every workplace.

The mere mention of the words "outsourcing" or "off-shoring" causes knuckles to whiten, teeth to clench, and veins on temples to bulge. This is because people interpret these words to mean "jobs being stolen." In gaining an understanding of what outsourcing and off-shoring really are, it is necessary to separate the two terms, even though they are often used interchangeably.

Outsourcing is a process whereby a job once performed within an organization is performed outside the organization. Off-shoring, on the other hand, is taking a job in one geographic area and relocating it somewhere else, usually to another country[1]. An organization that off-shores is not necessarily engaged in outsourcing; for example, the use of different divisions of the same company. An organization that outsources doesn't necessarily engage in off-shoring; for example, using a contractor instead of doing the work in-house.

Most of what we hear concerning outsourcing focuses on the off-shoring element --- U.S. companies sending jobs to China and India --- rather than which functions of the organization are relocated. But as rising salaries begin to lift the living standards in other countries, the attraction of "cheap" labor begins to fade. Such improved living standards, however, are not country-wide, and vast largely uneducated underclasses still exist who receive little if any of the benefits of foreign investment. India in particular is constrained by a lack of infrastructure to support companies wishing to establish themselves there.

Beyond China and India, Eastern Europe and Russia are becoming the next front in outsourcing. Hungary, Poland and Russia (to name but a few) possess high degrees of technical talent advantageous for high-skilled industries. For example, Boeing established a center in Moscow in the early 1990s to utilize some of Russia's highly skilled engineers in their commercial aircraft development[2].

It may surprise you to learn that the U.S. is also a recipient of off-shoring. Although media attention focuses on the reverse, large foreign organizations commonly establish U.S. offices, which is a form of off-shoring.

So what does off-shoring do? In addition to providing a jobs boost to the economy the jobs are off-shored to, it's usual for those jobs to have higher wages than comparable jobs within that economy[3]. News reports would have you believe that jobs being outsourced to India and China represent massive reassignments, causing serious deterioration in the market from which they are taken. While there is an element of truth to this, the number of jobs classed as "lost" are outweighed by those created through new developments. The problem occurs when the process of job creation is looked at under a microscope. Job creation is not a homogeneous process, and while the numbers may look good on paper, there can be large increases and net losses from one local region to the next.

Most jobs have disappeared from the manufacturing sector largely due to its commoditisation. Productivity gains resulting from automation and reduced costs through the utilization of low skilled labor overseas has eroded this sector in many developed countries. This creates the problem of how to re-engage and retrain that freed human capital for jobs within the new economy. Part of the solution comes down to individual choice; will these individuals cast their gaze forward, retrain and try to reinsert themselves into the market or will they continue to look back for the same job?

One mainstream media column noted: "Despite all the publicity in the United States about jobs being lost to India and China, the size of the IT employment market in the United States today is higher than it was at the height of the dot.com boom."[4] Illustrating the issue of geographic disparity, many midwest states may not share in these new economy jobs because their workforces are simply inappropriately trained to take advantage of them. Hence, despite surplus human capital, companies in Silicon Valley may continue to carry vacancies for months at a time while others remain unemployed unable to access such opportunity.

In 2005, Thomas Friedman, Pulitzer Prize winning journalist for the *New York Times,* provided an address to the U.S. National Governors' Association meeting held in Des Moines, Iowa. He had just published his now famous book, *The World is Flat*, the title referring to the globalization phenomenon and the levelling of the playing field between countries such as India, China and the rest of the developed world. During his address he spoke on this issue of skill matching and the needed flexibility. He advised students to pick up early on the actual process of learning, and that the greatest skill they could possess was to "learn how to learn."[5]

If you know how to learn, you can constantly re-invent yourself and respond to the needs of the market. "Learning how to learn" means you can roll with the waves of change and remain dynamic in an ever-changing environment, whereas stopping after completion of formal education means the incessant march of time will gradually erode your skills until they become irrelevant to the shifting marketplace. This is the situation facing many people in this most recent wave of layoffs; unemployment due to a structural shift in the economy.

There are other reasons besides labor costs for companies to off-shore. One is to escape the crushing and in some cases debilitating social contracts they have with employees --- that is, the extensive benefits packages bestowed upon permanent employees in the form of medical and dental benefits.

Ubiquitous health coverage is not part of the national infrastructure provided by the government. Off-shoring to other countries not encumbered with footing the bill on such costs represents a large saving to the parent organization. The generosity of pensions and benefits packages given to employees have started to bite organizations that offer them, in many cases becoming too much of a burden to bear. In these cases, filing bankruptcy is the only way some have been able to remain afloat and escape these large burdens. Examples such as Delta in the commercial airline industry and General Motors in the automobile industry have their origins in trying to escape the social contract.

We have all seen reports of call centers established in India and manufacturing plants shut down and sourced to companies in China. Politicians and citizens scream out to these companies to retain local jobs, but it is a losing battle. Why? Do companies not care? Whether they care or not, they are now competing on an international scale as opposed to the local pre-globalization markets. With the tearing down of trade barriers between nations, companies feel the cold wind of international competition. In the local market, there may have been competition with two or three other providers; in the international market, there may now be twenty or thirty.

A pseudo-outsourcing that some are more comfortable with occurs between countries with historical connections. In the U.K. for example, the building and construction industries experience a constant imbalance between the supply and demand of engineers. Long standing Commonwealth ties with countries such as Australia means that many young engineering graduates (particularly those from mechanical and civil engineering disciplines) from that country usually find no trouble obtaining a position in the U.K.

The other side of the coin, however, is labor regulation (such as in the U.S.) which is emerging as a big impediment to the free movement of high-tech employees like engineers. Those intending to work in the U.S. must have, at the very least, an H-1B visa, a short term work permit that many employers in the Information Technology industry have used to great effect during the dot com boom. While the dot com boom has come and gone, the number of H-1B visas is still quickly depleted after their release by the federal government. This has led to a rising backlash by various industry groups in the U.S. against the visa's prolific use and alleged abuse. Many industry groups (such as the IEEE) complain that using H-1Bs encourages employers to look off-shore for technical talent, and to pay them lower wages while in the U.S. due to the labor rate disparity that usually exists between the U.S. and their country of origin.

Those coming to the U.S. from the Middle East and China experience even further layers of regulation due to measures put in place since the 9/11 attacks. It can take many months or even years to obtain a visa, which has led many people who would have previously focused on the U.S. to now look to other countries, such as Ireland and Canada, which have more accepting attitudes. With the ever-increasing global shortage of talent, the international mobility for people with the right skill is more likely to be catered for.

Despite the ability of globalization to bring developing countries into the developed world, some argue that off-shoring is not all it's cracked up to be. While on a business trip in Europe, I met a gentleman who worked for an electronics manufacturer in the U.S. He told me that their company's experience with off-shoring was not positive, primarily because their product had to be manufactured with a high degree of quality, something that a focus on low-cost, low-skilled labor cannot provide.

Some employers have tried to "have their cake and eat it too" by resorting to "localized outsourcing." Instead of outsourcing to China or India, there has been a move to outsource to other parts of the U.S. where wage rates are much lower. While they obtain some of the benefits of outsourcing while keeping the jobs inside the country, they must still adhere to the social contract many employers are bound to by permanent employment. As a response, there has been a move towards greater casualisation of positions, that is, removing the social contract applicable only to permanent employees.

Any position that can be defined, categorized and boxed --- so that a manager can say "that position there does this and only this" --- will be fair game for outsourcing. I'm sure a lot of people who have had their position outsourced would say: "Hang on a minute, there's a lot of skill and creativity that I put into the tasks I do, you can't outsource that!," which is undoubtedly true. But ask yourself this: does the position REQUIRE it? If not, and it's just an extra feature you add on to the task, then at the end of the day, in the cold hard light of how many extra units it ships or extra defects it catches, if the answer is zero, then unfortunately the market doesn't view things the way you do. It's all about the value-added proposition.

So if those engineering jobs that are easily characterized and defined are fair game for off-shoring, the next question is: what jobs aren't? Two groups of positions that will never be off-shored from the U.S. are those related to national security and defense, and those that cannot be neatly packaged.

I don't see how national security positions will ever be off-shored. The whole idea of placing positions that deal with the security of the nation into the hands of workers in other countries is nonsensical. Such positions usually require security clearances that involve lengthy background checks.

A second set of positions I doubt will be off-shored are those where the engineering cannot be commoditised or easily replicated elsewhere. For example, if you say to someone: "Come up with a software tool that gives me these outputs," then that could be off-shored anywhere there are coders. However, if you need an engineer to work with a client organization in need of a partner in developing a solution, rather than having it pre-defined, then you need someone locally to interact with the client face to face.

Another way to view the situation: just assume everything can be outsourced, and rely on your ability to move quickly. With this kind of perspective in mind, those engineers who are flexible and adept at a systems level will be able to quickly fill any vacant chair when they need to, rather than only being able to fit into one or two defined and static roles.

Move to a services perspective

One way engineers can save themselves from being commoditized is to push themselves further up the value chain. You may have seen the term "value added" used with respect to companies operating according to the rules of the new world economy. This term simply refers to a company that, by virtue of its involvement, has "added value" in some way to a product or service. While they don't create the products consumers purchase, they have manipulated the product in some way that enhances its original value to the consumer.

A good example of a company that operates this way, and explicitly acknowledges so in their marketing and advertising, is the chemical company BASF. In a 2005 television ad campaign, the catch phrase repeated over and over was:

We don't make a lot of the products you buy, we make a lot of the products you buy better.

Another big organization also in the value added sector is Toyota. You may think, "Hang on a minute, doesn't Toyota make cars?" And you'd be right; but Toyota sees itself in a different light, and if you think it doesn't matter how a company views itself, note that Toyota is the only car maker in the U.S. that has not closed down a single plant --- and they're actually looking to open two new ones. Featured in Fast Company's Dec 06/Jan 07 edition, Toyota is focused "not on how to make cars, but how to make cars better."[6] This is an example of a company raising itself up the value stream, from one that makes cars, to one making better cars.

"Value added" often gets discussed when comparing manufacturing or agriculture to services-based industries such as accounting. Australia's economy, for example, has traditionally been based on the resource and farming sectors. It's what they're known for around the world, with the image of the outback and the drover exploited for many an international tourism campaign. Despite the fact that Australia is a developed nation, 80 percent of its 2004 exports were resource-based, primarily from the farming, minerals and mining sectors[7].

The problem with such a heavy reliance on resources is that they are finite, and therefore will eventually dwindle to the point where continued extraction is no longer viable. If all you do is sell ore to other countries, leaving it to them to add value, there's not much else you can do with the revenue stream other than exploit it until its exhaustion. Multiple revenue streams and disproportionate returns come from moving beyond mere export and actually doing something with the ore. At that point, you are moving towards extracting revenue not just from the tangible material but also the value added services on top of that.

The services sector has been the fastest growing sector for a number of years, and the fastest growing element within it is "value added" services. Why is this the case? Providing a service has many advantages over the production of tangible goods. They don't use resources; if not used, services just disappear. For example, not going to your accountant to get your tax return doesn't mean they just sit in their office doing nothing for the hour it would have taken them to do your return, they do something else. The ability to reallocate their efforts based on the situation is infinite. What this translates to for engineering is that the entire lifecycle of a product, not just the product itself, is a source of revenue: maintenance, providing replacements and upgrades, transitioning out of service, disposal and planning for successors. As you can appreciate, such an approach relies a lot more on understanding what a customer's objectives and goals are, rather than simply providing them with a product and then washing your hands of it as soon as it's paid for.

Legacy Systems

In our incessant push towards modernization and staying current (especially in western nations), we need to remember that we don't live in a completely disposable world. We don't replace a building that's a few years old just because new architectural styles roll in or new methods of construction are developed. Despite new advances in water sanitation and purification, the water authority doesn't build new sanitation and treatment plants every five minutes. Large capital projects are expensive undertakings that can take many years to complete and must be capable of meeting demands well into the future. The fact that large amounts of public funding has been sunk into their development highlights the obvious fact that the maximum return on investment needs to be realized in order for the public to have confidence their taxes are being utilized effectively.

Therefore, while engineers need to be aware of future technologies and their application to solving present day problems, we cannot do this at the expense of those that have already been developed. What needs to be considered as part of any new design is just how it is to interact with older legacy systems. We've all tried to open an older computer file with a later version of the same software, such as trying to use Word 98 files with Word 2003. While there may be some functionality, there's no explicit design to ensure compatibility between the packages.

> **Compatibility with legacy systems at Airbus**
>
> A high profile case of just such a problem was experienced with Airbus and the software it used for the design and development of its A380 Superjumbo. Different Airbus centers had been utilizing different versions of the same and different software packages which caused a number of problems[8]. Having pushed the problem aside for too long, Airbus faced delays in getting the A380 to its customers --- this issue of legacy systems was one of the prime reasons why.

Legacy systems also play prominent roles when dealing with developing countries. Existing water, sanitation, power and other civil infrastructure, while antiquated by western standards, needs to be integrated when trying to improve facilities for these countries.

Global mobility

Many young professionals around the world seek experience working in foreign countries, often aspiring to make a permanent move abroad in search of better prospects. Engineers exist in every corner of the world, so wherever you go there will be the potential to obtain employment as an engineer. This wasn't always the case. Even as few as ten years ago getting a job as an engineer overseas was an unusual occurrence. This changed with the introduction of the internet, which enabled instant global communication in ways the telephone never could, and the globalization of the workforce, brought about by the rationalization of the 1990s and the deregulation of markets. Such global hopping is now much more common, but there are still obstructions in the process.

What do you need to become one of these global hopping engineers? First, you need your engineering degree to be recognized around the world. Secondly, you need your home country's engineering licensing requirements to be recognized in the new country, or be able to work towards certification in the country you're migrating to. Finally, you want to know what is happening in your areas of interest in different parts of the world.

The movement towards standardized degrees or recognition has begun in a limited way within Europe. The Sorbonne Declaration, signed by France, Germany, the U.K. and Italy in May 2000, harmonized and matched educational systems resulting in just three levels of degrees: bachelors, masters and doctorate. In 2001, 25 more countries from the European Union signed the agreement, now known as the Bologna Declaration[9].

Similar to this is the Washington Accord, signed in 1989. This agreement exists between signatories who recognize the substantial equivalency of their respective engineering programs. The signatories include Australia, Canada, Hong Kong, Ireland, New Zealand, South Africa, United Kingdom and the United States.

While such accords and agreements allow qualifications to be recognized around the world, students still tend to internationalize themselves after graduation. What do I mean by this? While some students elect to study for an undergraduate degree overseas, most international students undertake a masters or some other graduate degree rather than undergraduate. One cannot start out with a global focus, it's necessary to start local. This seems reasonable --- how can you jump to the global stage when you don't even understand engineering and the issues surrounding it in your own backyard? This is where the idea of an international institution becomes relevant.

The 7th WFEO Congress on Engineering Education, held in Budapest, Hungary in 2006, focused on the "Mobility of Engineers" and the various elements required to enable it. The idea of a "World University of Technology" where engineers can go to learn about engineering issues on the world stage has been around for some time, and was raised again at this conference. The Bologna Declaration and Washington Accord are two examples of recognition and equivalency across borders --- but how ready are we to have such an institution?

I've seen how such an institution can work while attending the International Space University (ISU) in Strasbourg, France to complete my Masters degree. This university takes in students from all over the world for a year at a time to complete Masters degrees in either Space Studies or Space Management. It is highly regarded within the global space community; NASA, the European Space Agency and other public and commercial institutions have sent numerous members of their workforce there.

ISU's guiding philosophy is the Three I's; International, Interdisciplinary, Intercultural. Knowledge is imparted to students with the constant reinforcement that the issues they deal with have a global impact, that they are not just technical in nature, and not from one religious or political ideology or another. Graduating from this institution was pivotal in preparing me for professional involvement in the international community.

While agreements such as the Washington Accord [10] seek to allow engineers to practice freely in any of the countries party to the agreement, international mobility for engineers still has a ways to go. Ideally, the market forces of supply and demand would determine where engineers went; but alas, this is not yet the case. Labor markets and their regulation is usually a federal responsibility, and the provision of the right employment conditions is something no government would dare scrap in order to make their country's employment market more open. Hence, while a young Australian can travel to Europe and get a summer job waiting tables, it's a lot more regulated --- and specific work permits are needed --- if you want to work as an engineer or other professional without a European passport.

What would be the result of freeing up the world's labor markets and enabling a global workforce? This was the focus of a World Bank study, the results of which are contained in the report *Global Economic Prospects 2006*. Somewhat surprisingly, the report revealed that were such a regime to be put in place, the "free movement of labor across the world would double global incomes."[11] Part of this rise in income would be caused by "remittances," foreign workers sending some of their income home. While there are positives and negatives from this practice, there seems to be a net positive through the reduction of "the incidence and severity of poverty."[12]

So, some countries are beginning to recognize that such tight regulation of their labor market does nothing to protect local jobs, and can destroy their attractiveness to the best and brightest in the high-tech fields. Germany, for example, possesses a tightly regulated labor market causing areas such as R&D to have suffered greatly --- so much so that Germany has in the last few years embarked on a plan to attract skilled migrants through competitive salaries and the lessening of bureaucratic red tape.

Next door, France was shaken in mid-2006 by extreme protests from student groups demonstrating against the plans of the de Villepan government to make it easier for employers to hire and fire staff. Such labor market reforms are designed to free up the employment situation so employers can move quickly to obtain and place appropriately skilled employees where they are needed. Similar reforms were passed in Australia --- but there is a downside. While such reforms make an employer more nimble on their feet, they also create a perilous new environment for workers, the most vulnerable of which tend to be those entering the workplace for the first time, such as recent school and university graduates.

> **Global embrace --- Ireland**
>
> A somewhat more positive example has been Ireland. Not typically considered a country with a large high-tech workforce (due to its small size, and perhaps its association with Guinness more than anything else), Ireland has surged ahead thanks to deliberate government focus. The Irish Research Council for Science, Engineering and Technology (IRCSET) announced new funding of almost $5 million to attract up to 50 postdoctoral researchers in the sciences, engineering and technology. The only criteria a postdoctoral researcher must meet in order to be awarded funding is their intention to undertake their first position at an Irish research institute the following October[13]. In addition, in 2005, Ireland's national science body (Science Foundation Ireland [SFI]) committed $11.7 million to develop the Irish Software Engineering Research Centre (ISERC) at the University of Limerick[14]. This has already been instrumental in many Silicon Valley icons, such as Google and Seagate, establishing a presence in Ireland. It has now become "far and away the primary location for the digital media industry in Europe, and second only to Silicon Valley in the U.S."[15]

Such results show how a clear government focus --- along with funds to reinforce intent --- can enhance global competitiveness, signal attractiveness to foreign investors, and encourage young engineers and scientists to relocate. After all, talent will naturally migrate to areas where conditions are most favorable, and in this case the old adage "build it and they will come" certainly holds true for young workers seeking the greenest pastures.

Lifelong Learning

If you are to become a part of this highly skilled, in-demand pool, you must embrace the concept of "lifelong learning." Perhaps you've heard this term but aren't clear about what it means. In the past, information didn't change much and change was predictable, linear and incremental. Once in a while something came along that created a bit of a bump in the road, but pretty much these things could be catered to. Employment was long term, conditions were stable, and money typically wasn't something you could negotiate about.

Come the late 1980s and early 1990s, and all hell broke loose. For a number of reasons companies began downsizing in response to the opening up of previous government monopolies: commercial entities supposedly would run things more efficiently, competition among providers would keep costs to consumers down, and consumers wanted greater choice. Governments saw that if these services were at least partly privatized they could reap the rewards by selling off some of the assets.

In line with this, the development of technology began to accelerate. The amount of new information and the rate at which we were acquiring it skyrocketed. Not only that, but this new information proved that some of our ideas were not entirely true. "I'm bothered by the ephemeral nature of information today. Way down in my stomach I wish that my assimilated knowledge would sit still," stated Robert Lucky, former head of Telecordia, hinting at the perishability of knowledge in an IEEE Spectrum magazine article.[16] Most of us think once something is discovered, it's been discovered --- but because it is a human endeavor, results can be open to interpretation and old theories based on less than complete information fall by the wayside as they make way for updated versions driven by better or more complete data.

The rapid pace of computer development is largely responsible for this phenomenon. Continual increases in processing power coupled with reductions in price made computer prices plummet in the 1990s, in turn greatly reducing the cost to develop new applications that streamlined business processes. Organizations became fixated with how much they could lower the bottom line. The workforce, instead of being seen as the key enabler of an organization, was now viewed as overhead, a cost to be minimized --- which they were as large numbers of employees throughout various industries were cut in massive layoffs.

This was the beginning of the end of "jobs for life." Gone was the idea of finishing university, getting a job, and keeping it until retirement. Not only would you have to prove your worthiness to a company to get your first job, now you'd have to prove it multiple times at multiple job interviews throughout your career. So, such job jumping has made "lifelong learning" a necessity for anyone who wants to stay current to the workplace.

Employees can no longer rest on their laurels. Their status at an organization depends on how they can add value to it, which is solely dependent on the skills and attributes that person possesses. Given the rationalization and globalization of the workforce, the responsibility for lifelong learning has shifted from the employer to the individual. Each person is now effectively their own CEO; *you* are the product *you* are trying to sell to potential employers. At the same time this responsibility has transitioned from employer to employee, the employer's focus has also shifted. No longer able to gaze out five or ten years and predict the future, their shorter horizons force them to exhibit much greater flexibility than they did before.[17]

We are currently seeing the results of such short-termism. A 2005 issue of *Fortune* magazine featured a cover story on top executives who had been laid off. Many had been made redundant due to their unwillingness or inability to keep pace with technology or developments in their field. This demonstrates how not paying attention to maintaining ones skills maintenance eventually brings them to a point of irrelevance to the job market; too far behind for anyone to invest in bringing you up to speed.

So the next question is, "If I need to update my skills, which ones should I focus on?" There's a couple of issues here. First, are there skills that you need to keep current related to your particular endeavor of engineering? And second, is there a need to hone skills that are more general in nature? Job-specific technical skills include things like learning the latest programming language, or keeping up to date on the latest techniques in your area. These skills, unfortunately, depending on how rapidly your organization's environment changes, are the most likely to decay rapidly. The general skills to keep updated include written and verbal communication, and networking. These are the least likely to decay as they are transferable no matter what job you go into. Networking is especially likely to allow you to choose where and when you move into your next position. I'll discuss this more in the section on networking.

Office politics

As engineers we tend to think our work will speak for us, and that we'll stay out of that nasty game that tends to be the preserve of the non-technical worker. We think our boss will see who is truly valuable, as if there's some kind of definitive just result. Alas, this is not always the case. We've all seen it: the one who panders to the boss or takes claim for someone else's work gets that position or plum role ahead of you. You may ask yourself: "Why? Isn't the world just? Won't good always triumph over evil?" These things happen because our bosses are human, just like us, with egos to stroke, and their own perceptions and biases. Because of this, they won't make decisions that are right or just all of the time.

As people of logic, we engineers tend to want to have all the facts before arguing a point. Sometimes, however, in order to remain visible, we've got to go with an incomplete picture. While we might see this as imprecise or sloppy, if we don't speak up at these times the boss may think we're not interested or have nothing to contribute, and therefore may not consider us for the more senior roles that those involved with the politics will be considered for. There is something to be said for shooting from the hip or going with your gut.

Let's identify what we mean when we use the term "politics." Most people think of politics with a negative connotation --- breaking promises, being generally underhanded, etc. Simply put, however, politics is nothing more than a negotiation between two or more people each with their own agenda and seeking to influence the outcome to meet their needs. Believe it or not, politics has a crucial role to play in everyday working life; in fact, work could not move forward at all without politics.

In an ideal world, companies would have all the resources they needed to carry out every project proposed by their workforce. Project Managers and engineers would merely have to ask for money to fund their project and it would be there, without any need to compete for it against other similar projects. Such an idealized view of the organization is commonly referred to in industrial circles as the "unitarist" perspective,[18] which essentially asserts that everyone is on the same team and they all want the same thing. As I'm sure you'll appreciate, this reality is quite rare.

In the real world, organizational resources are limited (usually scarce), and projects compete among one another in order to obtain funding. Each group thinks their project is the most important, and there can be a difference of opinion on which project is best for, or speaks to the goals of, the organization. The primary determinant for funding rests on the strength of the case the Project Manager or lead engineer presents to those with fiduciary responsibility, usually senior management. Therefore, any funding request must be presented in terms management is most familiar with (usually terms addressing business and customer need), and in so doing we arrive precisely at the operational definition of politics: presenting an argument to influence an outcome.

Workforce flexibility

The great country and western singer Dolly Parton released a song in the 1980s called "9 to 5," later turned into a movie of the same name. This song lamented the eight-hour grind, with one of the lines: "It's enough to make you crazy." When it comes to the hours we work, things haven't changed a whole lot.

There is a lot of rhetoric from companies when it comes to work/life balance. Their values sheet will tell you that work/life balance is important to maintain, but look a little deeper and you'll usually find no concrete policies backing these statements, or worse, policies that are not enacted in the workplace. When it comes down to it, a 40-hour work week is still standard (a notable exception being France's 35-hour week) and in some cases is seen as the bare minimum; often those who work more than 40 hours do so with a certain sense of pride, something they wear with a badge of honor.

Sometimes it's going to be necessary to work a little overtime for whatever reason, but policies that focus on the amount of time at work rather than actual productive output create a topsy-turvy world that prizes attendance over value. "I worked a 10-hour day last Thursday" makes one feel somehow more valuable to the company, although nothing could be further from the truth --- it's a stereotype that needs changing. As the current head of a NASA field center once said to me: "If you can't get everything done in 8 hours, you're obviously not doing something right."[19]

As previously discussed, a company's main engine for output is its people. Looking at the company from a scientific perspective, the equation for its prosperity is very simple: people are the only asset a company has at its disposal to do anything! This has been recognized in Japan which has reaped the benefits from the teachings of Peter Drucker, the iconic strategy and management thinker. Japanese employees are looked upon as the company's most valuable "resources." They are considered long term employees and treated as part of an extended company family. Compare that to the U.S. and other western nations where employees are considered primarily in terms of their "costs." This is reflected in the move away from permanent employment and career jobs to temporary and contract positions.

If your people are not operating at their optimal levels then you're going to get sub-optimal performance. This is a simple statement that would draw little disagreement. In order to provide optimal performance we need to know its causes. To find out, we need to introduce the ideas of Abraham Maslow and his famous *Hierarchy of Needs*.

If you work for a company that pays you a wage, it will seem like you are working for them -- but ultimately, you're working for yourself. While you receive money in exchange for employment to pay for your subsistence and luxuries, who you work for is at your discretion; therefore, you are ultimately in control. This is not the position most people see themselves in --- in fact, quite the opposite. Statements like, "I've got to go to work today," "I wish I could quit my job," and the like reveal the utter lack of control we feel.

The only way we have to assert some kind of control over our working lives, other than resisting the toil of work, is to head in the opposite direction and work to excess. Many work long hours not because they have to but because their identity is so tied up with their job that to not be at work is to lose a sense of one's self. "We have reached a point where hours that were once seen as exploitive are now passed off as a lifestyle choice."[20] Many men still feel the traditional need to be the primary source of income in the family, as this is how they've been conditioned.

Those of us who are salaried employees are not working more hours due to deep-seated loyalty to the company (although those who have been in the workforce longer may feel this way), we're working longer hours due to the fierce competitiveness with our co-workers --- we want to outshine them. Many of us need that context to make the work more interesting or exciting. Engineers tend to be intensely competitive but, as mentioned earlier, it's a fallacy to think we're in a long term competition with anyone but ourselves; therefore such gestures of beating the other person are ultimately futile.

The underlying reason behind such action can best be summed up with the term "Status Anxiety," made popular by the modern day Swiss philosopher Alain de Botton. In his book of the same name, he deals with the issue of what other people think of us, and its natural extension: whether we ultimately feel like a winner or a loser[21].

One of the strongest motivators of status anxiety is the accrual of financial wealth, particularly predominant in western nations. Such action, in fact, is an attempt to find meaning or fulfilment not being met through other means. As engineers, this meaning or fulfilment can often be realized through the constant solving of problems, or feeling a sense of contribution to a project or product. Forgetting this and focusing only on remuneration leads to disappointment. The 19th century German philosopher Arthur Schopenhauer said it best when talking about the measure of success: "Who is successful and who is a failure is based on nothing more than suspicion, rumor and fashion."[22].

Creativity and Innovation

As I mentioned earlier in this book, there are economic advantages for a country to possess a technically literate workforce. With current discussion focusing on the "greying" of the workforce, it's worth noting that technical literacy coupled with creativity is not age dependent. Unlike soccer stars or ballet dancers whose careers rarely last a decade, engineers continue as long as they have their faculties about them. With more and more people working past retirement, it would be nice not to be forced to switch professions at age 65. Paradoxically, age does have its advantages in the engineering world.

Unlike their older more savvy colleagues, young engineers can face a number of barriers when it comes to creative invention --- particularly when it's time to get technical ideas into the marketplace.

The inventions or ideas of young engineers can quickly be overtaken by those with greater experience and/or ulterior motives. Recognizing and avoiding such exploitation is not generally taught to the young engineer, but is often learned through cold, hard experience.

When it comes time to invest in a new product's development, the inexperience of those with no track record can loom as a large stumbling block. Only those inventors with a PhD, formal accreditation, or at least some background or career history will provide a feeling of reassurance to the investor. For this reason young inventors face an uphill battle when trying to obtain funding for ideas that, presented by more senior engineers, would likely attract investment.[23]

As you can see, creativity and innovation are qualities not necessarily impacted by education, age or experience; they are intangible, innate qualities that people either have or don't. You can't quantify the level of a person's creativity in a test the same way you can grade their mastery of the English language, and in turn compare them with others. "John is more creative than Mark" cannot be determined in the same way as "John is a better speller that Mark."

It is puzzling, therefore, why in the modern workplace where people work as many as 70 hour weeks there is an expectation that their creativity will extend proportionately. Positions relying on creativity and innovation lend themselves to shorter working periods, typified by frenzied creative storms followed by quiet dormant periods that allow time for reflection. Dynamism followed by dormancy is needed because it allows one to recharge and get ready for their next creative burst. Unfortunately, it is the reflection time that often gets pushed into the background.

Many engineers are under constant pressure to get things done or deal with ever present crises, all of which cut into this reflection time. Recognizing the rewards that come from reflection time, Google has a policy of allowing 20 percent, or one day in five, for employees to pursue their own ideas. This policy has resulted in some of Google's most innovative and profitable ideas, and they have reaped the benefits. It's good policy for companies to understand "that productivity and creativity don't necessarily flow between the hours of 8 am and 6 pm."[23]

It is just this view of a standard work week or standard hours that cause those women who temporarily exit the workforce to raise a family to never come back. The notions we have of their utility is wrapped up in our "working day" mentality. If we're trying to encourage what few women there are in the engineering workforce to stay, discouraging them in this way is counterproductive.

Collaborative culture; the new normal

The way young people collaborate today resembles a networked or, to borrow a term from the internet, "peer to peer" fashion. This trend is also reflected in the video game culture that forms such a large part of the lives of teens and twenty-somethings. Games such as the "Sims" and "Star Wars Galaxies," unlike those of the past that were individual pursuits, present players with more than just the linear win or lose outcomes. You "experience" modern games. Winning is now defined by the amount of time your character is in the game. Contrast this with the objectives of past games, where eating dots (Pac-Man) or shooting invaders (Space Invaders) or blowing away bad guys (Doom) was the definition of victory.

Game makers have extended this experience for players by allowing them to connect with their friends, making the internet an integral part of gaming. Groups of players form teams that compete against one another as online virtual combatants, such as in Star Wars Battlefield or any of the hugely successful Half Life first person shooter series.

Working towards a common objective with others develops valuable teamwork skills in young game players. They are also forced to contend with multi-dimensional complex situations, much the way engineers must consider the entire lifecycle of a product rather than just focus on its initial design when starting out.

Consider a game like Command and Conquer, where players balance spending money to train troops with the need to develop other support facilities such as bases to house them, power stations to meet increased demand, etc. By dealing with such problems, players learn about the important subject of logistics support which every international company deals with as they grapple with the transition to global supply chains.

These games have become so successful and their potential to develop problem solving has been recognized as so great that they are now being used by the U.S. Army and Marine Corps. Even before they have walked into a recruiting center, these games familiarize potential recruits with concepts the armed services would expose them to. This means when recruits arrive at boot camp, they are already familiar with some of the ideas they will be exposed to.

It's not just the military that's utilizing video games. Even religion is in on the act with a first person game called Catechuman --- in which instead of shooting people players banish evil spirits and rescue captured brethren. Instead of trying to ram religion down the throats of young people, the game's creators have found it far more effective to embed their message in video games where today's youth spend most of their time.

Such an "information push" has been experienced by anyone who has received an email by being on a distribution list. While email's use has greatly increased connectivity and accessibility, it hasn't led to increased responsiveness --- consider the fact that half of the email we receive is either spam or been unnecessarily cc'd. Email has established a many-to-many connectivity which, in some cases, can be downright annoying, as anyone who's tried to conduct configuration management with 15 different versions of a document floating around in cyberspace knows.

One solution, the "wiki," emerged from the Open Source movement. "Wiki" is Hawaiian for "fast," and it acts like a central webpage or repository. Containing a raft of different topics, wikis are a bottom-up style of information development. Previous editions are retained so one can roll back or discard changes that have been made. Among its best features, wikis allow real time collaboration on the same document, like having a virtual whiteboard. Wikipedia, the most well known wiki, is an online encyclopedia continually updated by thousands of anonymous users.

The use of a wiki represents an "information pull," where entries are drawn from a number of different sources as opposed to an "information push" like a television broadcast. The old saying "many hands make light work" is the kind of advantageous task division enabled by a wiki, whereas "too many cooks spoil the broth" represents the confusion and extra complexity often introduced by having too many people collaborated with through email.

The big drawback to wikis, however, is that they reside on the internet and therefore can be accessed by anyone. A related issue involves the legitimacy and accuracy of the information --- it is only as good as the ability of those in the network to distinguish right from wrong. With corporate entities increasingly concerned about security, they are understandably anxious about web enabling information unless safeguards such as corporate firewalls or corporate intranets are in place. Unfortunately, knowing when enough security is enough is usually retrospective, measured after a cyber-attack or other security vulnerability has been spotted.

Use of these networked tools and online experiences affords engineers the prospect of developing more holistic appreciations by being able to reach across geographical divides to source information wherever it resides. However, increased virtual accessibility relies on the searcher to be more discerning; just because something is on the internet does not make it true. It also raises the concern that as engineers dive deeper into the virtual world, they may become increasingly isolated within their organization. This results in a collective of people who act effectively as a team in the virtual world but cannot interact with each other in real life. This has become a typical problem for engineers in the networking field.

Networking

We've all heard it many times: it's important to network with your peers, you need to network with those outside your specific area of expertise --- and you probably think that's a load of rubbish. "Why should I go to a seminar or event that's got nothing to do with my studies or my work? It's just a waste of my time." I'm sure I had similar thoughts while at university.

When it comes to networking, engineers initially think it's just a case of gathering a bunch of business cards from a group of strangers. Particularly among the younger crowd, many admit going to events because their boss "encouraged" them to --- an obligation, not really a suggestion.

But networking is not about amassing business cards or asking people for favors; it's about making connections with people from within and outside your industry or area of interest who can aid in expanding your knowledge base, whether that be about jobs, industries, companies or new technologies.

The best way to network is with a selfless attitude, not with the mindset of "what can I get out of this?" You should be looking to meet others to converse with, in order to understand what their needs are and, if possible, help them. Why should you help them? There is no compelling reason, other than a selfless sense of charity, and the realization that in the act of helping someone else, both the person who's being helped and the one doing the helping benefit. If you can provide help (and it doesn't have to be much) through a contact they are looking for, information on a particular topic, etc., then you are on your way to building a meaningful relationship with that person. More often than not, people will be willing to help you if you've provided assistance to them first. Do this a couple of hundred times and you'll have built up a network of people who are likely to not only put in a good word for you, but will also be far more willing to tell you about any opportunities they may hear about. In essence, networking increases the amount of active sensors you have out there in whatever environment you choose to consider.

"Netweaving" is the term applied to this kind of relationship building, and it is one of the best ways to make long lasting business relationships that have substance. Unlike flitting from one person to another making it apparent you couldn't care in the slightest about what the owners of your collected business cards have to say, netweaving is all about spending more time with less people and really finding out what they are about. It's more involved than just the delivery of an "elevator speech" or putting yourself in the best light to try and impress someone. It's listening to what the other person is saying and then responding in kind.

After establishing an initial rapport, any contact you have with that person in the future is more likely to be reciprocated, unlike the "let me know what I can do to help" often delivered on departing and used as a smokescreen to mask a socially awkward departure.

Ethics

Ethics is an issue discussed in university, but it's hard to get a good sense of it from a textbook or class. It's related to how your internal moral compass guides you. It's about doing things with integrity, honesty, and not being afraid to admit a mistake or a problem even though no one else may have seen it, been aware of it, or even care about it.

Ethics and making ethical decisions are on the minds of most CEOs today, given the speed which news of unethical decisions can be distributed to a wide audience through the media and quickly destroy an organization's image and credibility. If you need a current day example, look no further than the mess created by the phone tap scandal at Hewlett Packard.

Unethical actions give the offender a disturbing capability to inflict disproportionate damage to an organization because, long after the offender has moved on or been replaced, the stench of their unethical actions overshadows any good the organization may do. This is why the previously mentioned statistics indicating college graduates' lack of concern about cheating were worrisome.

I'll give you a quick personal example of ethical decision-making you may be able to relate to. I was parking my car at work one morning and slightly nudged the car in front of me. I got out and determined it was nothing too serious, just a dent in the car's license plate. The easiest choice would have been say that the damage "wasn't worth worrying about" and, as no one had seen me do it, walk away and think nothing more about it. The other option was to admit I had hit the car and do something about it.

I wrote a note to the owner of the vehicle, explaining the damage I had caused, that I was prepared to replace the license plate if necessary, and signed my name and phone number. I received a phone call later from the car's owner. He thanked me for offering to pay for a replacement license plate, but that it would not be necessary.

From this experience, I gained the knowledge that my own internal moral compass had been strengthened, that I had done the right thing, and that this had been re-affirmed by someone else as well.

This raises two points about ethical decision making. One: making an ethical decision usually takes more effort than an unethical one. In this case, it was easier for me to walk away than go through the trouble of writing a note and notifying the owner. And two: an ethical decision creates a self-reinforcing habit --- once you make an ethical decision, you're more likely to do so in the future.

Now it's not hard to do the right thing when the choice is between right and wrong, and many of you reading this will say that it's pretty easy to own up to something when the consequences are nothing more than a $30 license plate. The situation becomes significantly more complex, however, when you have to decide between a bad and worse situation. In other words, no matter what you do there's going to be some hurt involved. Therefore, most people expect they will ultimately be rewarded by their employer for doing the right thing, but this is not always the case.

> **Ethics at work**
>
> Take the case of Salvador Castro, a medical electronics engineer working for a company called Air Shields Inc. He found a design flaw in an infant incubator, and the company would not act on his concern. Castro was fired because he threatened to notify the Food and Drug Administration (FDA). He took his former employer to court for wrongful termination and has spent the last eight years in a messy court battle. Despite this, however, Castro said: "I'd do it again in a heartbeat."[25] This demonstrates that although he lost his job, he was coming out with his integrity intact.

It is not easy to make such a sacrifice to retain your principles. Imagine a worse scenario, however: fast forward a few years and the product gets widely distributed and finds its way into hospitals. A number of children die through the use of the incubators, and the FDA traces the deaths back to a design flaw. Mr. Castro had knowledge of the flaw, but his unwillingness to raise it as a concern would likely have seen him and the company face a whole string of civil and criminal charges. At the end of the day, the only person who has to live with the decisions you make is you.

Exchange of ideas

In the engineering community, we sometimes suffer from the "not invented here" syndrome. Our focus and dedication (usually a hallmark of engineers) can also be our Achilles heel, as it blinds us from considering other possibilities. In our haste to solve the problem, we forget to look around to see if this problem, or one like it, has already been solved, or if prior work affords us some amount of leverage.

An example of such tunnel vision is the inability of commercial entities to fully exploit the technology of the internet. Many software industry companies are racing to stay ahead of competitors in the capabilities they provide to market. However, when it comes to being on the cutting edge, those who reign supreme aren't even players in the software industry; it is the adult entertainment or porn industry. At a 2005 gathering of Financial Services CIOs from the U.K., one representative commented: "We're doing a lot of work on content provision for mobile devices and WAP access to online services, but we're nowhere near as far down the track as some of the porn companies are."[26]

This statement highlights the fact that collaboration, particularly with those who are outside your core market, can be far more preferable than trying to go it alone. The internet, initially created for the scientific community as a way of sharing information, has, in less than two decades, been the catalyst for an explosion of ideas and commerce. Many organizations grappling with how to deliver richer online content have done so largely in isolation from their peers; fear of giving competitors an edge has prevented collaboration or cross fertilization of ideas. This slowed down progress, and it points to the advantages engineers can bring to the table when they possess a view broader than their own industry or discipline.

Cross fertilization can come not only from within different sectors, but from outside of the business world altogether. Just look at how much the modern world has tried to gain leverage from biological systems and the millions of years of evolution these systems have had to solve environmental, biochemical and structural challenges. The point here is to be open to possibilities, no matter when and where they arise.

Sustainability

You've probably heard a lot about sustainability in the press of late, and wondered about terms like "sustainable business models," "closed loop processes" and "environmental harmony." Is this just a louder voice from the Greenpeace movement? And besides, what's this got to do with engineering and my early career as an engineer? Well, sustainability doesn't just concern the environment, and it's not just about stamping a green logo on your product or service --- it extends far more deeply than that.

Sustainability can be defined as the continuity of economic, social, institutional and environmental aspects of human society.[27] Placing this within the context of developing a product or service, it means looking to not only maximize the financial return to the organization, but extending it across other domains as well (public perception and environmental footprint being but two examples). An obvious dilemma arises when an organization identifies a course of action that will generate favorable financial returns but unfavorable consequences in these other domains (increased pollution as an example).

Despite the increasing focus placed on these other domains, it should be understood that sustainability is not the pursuit of a green agenda just for the sake of it; rather, it's the adoption of an integrated approach in the way an organization does business; it's not just about using less detergent when washing coffee cups in the break room, or about having separate bins for trash and recycling. An organization that truly has sustainability embedded within its culture makes it a part of every business decision, process and operational act at all levels. This is probably not something many organizations have achieved yet, as it requires a constant and never ending commitment to achieving a cultural change.

To further complicate the issue, moving towards sustainability and reducing unsustainability are not the same thing[28]. There are varying degrees of how organizations take on sustainability. The sustainability I mentioned above, making resource usage more efficient, is a version of sustainability that doesn't seek to alter the underlying process or fact that the resource is being used ---rather it uses resources at a reduced rate. The true intent of a sustainable perspective, however, is to stop business as usual, look at the underlying reasons for using that resource in the first place, and to question how else that activity can be done. From the use of recycled paper in printers, to the ability to alter processes in manufacturing so toxic emissions are reduced, to be truly effective, sustainability needs to be interwoven into the culture of an organization such that people are constantly questioning what they do everyday through this new lens.

Sustainability doesn't always have green overtones, although its usage in recent times has become synonymous with the environment. Ask any economist about "sustainable" profits and they'll describe the sweet spot businesses seek --- recurrent profits that can be generated into perpetuity. So when a business talks about sustainable practices, their terms of reference may differ markedly from those used by the general public. This highlights the need to be clear of the context when discussing sustainability.

Looking at the environmental side of sustainability, it's important to gain an appreciation for how resource intensive an endeavour is. When serving on the National Committee for Young Engineers in Australia, a fellow committee member developed an excellent representation of sustainability easily understood by the public. Based on a poster series she had developed called, "How many engineers does it take to make a chocolate bar," she identified all the resources required to make one chocolate bar --- and in so doing, outlined an environmental footprint for that product. Such resource maps make it possible to question whether so many resources are needed for a product's development and allow for the scrutiny of each step to see where it can be optimized.

Embracing sustainability meshes well with the product lifecycle. This concept is not new, but its incorporation into the sustainability argument has given it a new lease on life. The product lifecycle is an identification of the different stages in a product's life, from initial design through development, manufacture (if it's a tangible good) and eventual disposal. It's important for engineers to consider, since most of the time we get hung up on the front end, the design --- after all, that's where our talents are most explicitly called for (in a traditional sense anyway), and designing things is closest to what we learn at university.

Embracing a product lifecycle perspective requires one to see a bigger picture. In many cases, funds allocated for design and development are like the tip of an iceberg; they're the most obvious and visible part, but not necessarily the most significant. A majority of the funds go into in-service support, full rate production and maintenance. Therefore, as a young engineer involved in the initial design process, an awareness of the product lifecycle during your initial design might save a lot of heartache further down the road when someone has to support the product in a customer's workplace, or when it comes time to dispose of it.

A concrete example of sustainability and its integration into the design process can be seen in the latest commercial aircraft offering from Boeing: the 787 Dreamliner. This aircraft has been designed with sustainable practices and methods in mind, something Boeing has stressed in their press releases: "From initial design to the retirement of the airplane, we are seizing every opportunity to minimize the impact on the planet's natural environment."[29]

Viewing sustainability from another perspective raises the question of whether innovative engineering can be experienced in other parts of the lifecycle. When thinking of innovation, we tend to think of cutting edge technology; the fastest microprocessor or the highest strength alloy being two examples. But these endeavours are usually associated with the design aspect, while there are many other opportunities across the lifecycle to add value.

In the development phase, you're trying to create a system that is a physical manifestation of the requirements you've developed to meet the goals you've established. In this phase, you're focused on creating something for the first time. Once you've created the product, however, you then move into a different phase in which you shift your focus to leverage off of what's called the "learning effect." According to this principle, after you've done something at least once, repeating it many times will reduce the amount of resources expended to complete subsequent repeats. The learning effect may kick in relatively quickly if you're building 100,000 units of something (any kind of commodity like a nut or bolt), but it may materialize over a far greater time span if you're building extremely small quantities or one-offs, like a dedicated satellite for an interplanetary mission, or a bridge. In these low volume endeavours, it may take a person's whole career to reap the benefits of the learning effect.

Diversity more than just culture

Much has been made of the need to understand differing cultures in the workplace. Many organizations, however, have either paid lip service to this idea, not fully understood the effects it can have on the work environment, or have policies in place that are never cemented into the organizational culture, therefore never practiced. With the increasing international mobilization of highly-skilled engineers, it is quite common for teams to be comprised of two or three different nationalities. While such diversity may not present problems when those nationalities have similar outlooks and contexts (all western for example), it can become a major issue if outlooks differ and contexts are mixed. Add to this differences in language, and regional and provincial differences internal to a country, and it's no wonder culture can be a major stumbling block.

So just what is meant by context within the diversity topic? Broadly speaking, context refers to the degree to which what is said is meant and it can range from low to high. Low context resides in countries such as Australia, England and the U.S., to name but a few, where what is said tends to be what is meant and this, along with brevity and independent thought, tend to be highly prized. In the west, language is used to communicate in as direct a way as possible, speaking in what is analogous to the shortest path between two points, a straight line.

Contrast this with countries that are high context, such as China, Japan, Korea and countries in the Middle East. In these countries, the group is more important that the individual and saying what one means is rarely the case. Much communication comes in the form of non-verbal responses, and "saving face" is seen as very important. For countries in the Middle East, for example, language is not simply used for the verbal transmission of information. In Iran this concept is called *taarof,* and can be difficult to discern if one is not aware of it. "Symbolism and vagueness are inherent in our language," comments a political science professor from the University of Tehran.[30] Seen in this light, one can begin to appreciate why citizens of Western countries may perceive citizens in Middle East countries to be untrustworthy, and why those in Middle East countries are puzzled when Westerners do not pick up on the deeper connotations behind seemingly simple statements.

Our perspective on culture is not universal; it depends on where you were raised. For example, people originating from Eastern Europe or Russia may view American culture as very warm and accepting, as it contrasts with what can be perceived to be a rather harsh existence in their part of the world. Conversely, those coming from Spain or Italy view the American culture as very cold, because their cultures are family oriented, quite opposite to the individualistic and self-deterministic ways of the U.S. and the west in general[31].

An accurate understanding of other cultures can rarely be obtained from afar. A proper, holistic understanding requires a commitment to learn about that culture, study its language, visit the country, and engage with its people. I found this to be the case when working in France, and during the time I spent in Korea on a placement during my university days. My understanding of that country, incidentally, was augmented through my involvement in Taekwondo.

Being conversant in another language and understanding another country's culture can be extremely beneficial to an engineer, as most organizations today are seeking to expand their horizons. As you become a focal point within a team, such knowledge affords you an opportunity for growth independent of your technical expertise. It will likely afford you access to senior people within that group who seek opinions from those who've been to that part of the world before. Turning a blind eye to the impact culture has on business is done at one's own peril; business dealings can easily encounter serious road blocks due to a cultural faux pas.

Institutional Reform

As mentioned earlier, engineering is not homogeneous --- it is impacted by the challenges being faced in that particular region of the world. In Europe, the continuing move towards unification and integration is now being reflected in the education system. The Bologna agreement (as discussed previously) seeks to provide three levels of tertiary education so a degree of commonality will exist across Europe. Such standardization will make it easier for employers and employees to view the employment platform as European, and not its constituent countries.

Workplace reforms have also been pushed in Italy and France, designed to make those countries more competitive. The downside of such reform, however, has been the removal of certain safety nets for employees in government institutions. In France, for example, the de Villepin government was forced to back down from its workplace reforms by violent student protests opposing the new laws. Had they been instituted, employers would have had the ability to fire those who weren't performing in their first two years on the job. With the current workplace practices, however, it is nearly impossible to remove those who don't perform, a real disincentive for organizations to take on new workers without a proven track record.

While workplace reforms implemented in Australia sought to make the employment market more dynamic, they removed many of the previously entrenched workplace rights. Instituted to make Australia more competitive, many do not believe the potential human cost is justified. Australia's economy is the most prosperous it has been for nearly three decades thanks to a minerals and resources boom. This has created a high demand and high wages for engineers, currently in short supply. As other sectors, such as defence, are also seeking engineers, the country is looking for ways to encourage more students into science and engineering. Realizing that such measures will not take effect for years, the government's only short term solution is to entice skilled workers from other countries to the island continent, hoping to bridge the employment gap until the homegrown supply of graduates picks up.

As western countries lament students' falling interest in engineering, China is experiencing a surge of interest. This is reflected in the country's science and technology budget. In 2006, China increased its budget 20% over the previous year, to nearly 72 billion yuan or $9 billion. This increase forms part of a larger goal to double the per capita GDP of the country by 2020. Such investment is not seen as discretionary to the Chinese. When talking about China's need to retain a competitive advantage, which is beginning to weaken as living standards lift and wages increase, China's Minister for Science and Technology Xu Guanha commented: "We don't have a choice."[32].

> **China and Power**
>
> One of the people I interviewed for this book was Dr. Zhenyu Huang, a Chinese national who obtained his PhD in China majoring in Power Engineering. In recounting his reasons for entering engineering, and in particular Power Engineering, he mentioned: "I knew that I didn't want to sit behind a computer all day so software engineering was definitely out." Another major factor in his decision was the abundance of Power Engineering projects occurring throughout China, each of which had a very high profile. It is no wonder, then, that combining a high profile with generous government funding results in students flocking to be a part of engineering disciplines that will grant them the ability to be a part of such endeavors.

It should come as no surprise that engineering is granted such high status in a country with a population of over 1.3 billion people; every member of the communist party's top tier (those ruling the country) are engineers of one form or another.[33] In a country where the government is looked up to with respect, parents have little trouble persuading their sons and daughters to take up science and engineering. Just imagine the dinner conversation in China: "If you work hard as an engineer, you can become one of the leaders of the country." Where else can they say that about their engineers?

While China is producing sufficient numbers of engineers, they do not have enough engineers in the higher level positions, such as Chief Engineer or Chief Architect --- primarily because there is no legacy or history for companies to draw this talent from. Another problem is that Chinese engineers who go overseas to gain experience rarely return. One Chinese engineer who recently received his PhD in the U.K. told me he enjoyed living there because it was so much less crowded than where he used to live in China.

Chapter 8, if you only have 5 minutes...

- Outsourcing is taking a job initially done within an organization and having it done outside the organization.
- Off-shoring is taking a task or function conducted in one geographic area and doing it somewhere else, usually another country.
- Outsourcing does not necessarily mean off-shoring, and vice versa
- The U.S. also benefits from off-shoring, such as when foreign companies establish offices here.
- Some benefits created by off-shoring are jobs generation, wage appreciation, and an increase in living standards in comparison with the immediate environment before the jobs.
- Workers displaced due to off-shoring need to be re-engaged through re-training.
- Thomas Friedman, the Pulitzer prize winning author of *The World is Flat*, has commented that the greatest skill students could gain is to "learn how to learn."
- The social contract represents a large expense to an organization, one that is alleviated through off-shoring.
- Globalization has changed competition such that local markets are now accessible by any and all.
- In-sourcing involves sending jobs to other parts of the country where wages are lower.
- Positions most in danger of being outsourced are those that can be easily defined, categorized and bundled.
- National security and hard-to-package positions are ones that are not likely to be outsourced.
- A flexible approach with a systems perspective will enable one to move quickly regardless of the situation.

- Moving from pure provision of commodities to value-added services is another way for individuals and organizations to remain valuable.
- The services sector has been growing at a fast rate over the last couple of years.
- The need to integrate new solutions with pre-existing, legacy systems is particularly important in very large, expensive and capital intensive projects that last for many years, e.g. bridges, new buildings or new satellite systems.
- Most young professionals seek to work in another country.
- To be able to work elsewhere, your degree must be recognized outside of your home country, you must be able to seek certification as an engineer there, and you must be able to legally work there.
- A standardized degree and recognition is being sought within Europe, known as the Bologna declaration.
- The Washington Accord recognizes equivalency of engineering programs of Australia, Canada, Hong Kong, Ireland, New Zealand, South Africa, United Kingdom and the United States.
- It's usual for students to begin their international exposure after college graduation, but starting out in the international setting is gaining traction with the idea of a World Technology University being floated.
- Supply and Demand do not yet fully dictate the international labor market, but such an approach would double global incomes, according to the World Bank.
- Lifelong learning means a consistent and concerted effort to continued learning throughout an engineer's career.
- Responsibility for a person's career has shifted from employer to employee.
- The most transferrable skills to keep updated are communication and networking.

- Job specific, technical skills are the most rapid to decay.
- Politics is a negotiation between two people, each with a different agenda, seeking to influence the outcome to meet their needs.
- Politics is necessary as organizations do not have unlimited resources to pursue projects, and there must be some way to prioritize one project over another.
- A 40-hour work week is a typical minimum.
- For creative positions, short sharp bursts need to be followed by periods of dormancy and reflection.
- Maslow's hierarchy of needs shows that we are ultimately working for ourselves.
- Young engineers can face the problem of a perceived lack of experience when trying to get ideas to take root among senior colleagues.
- Creativity and innovation are difficult to quantify in a test.
- Team-work and a collaborative culture are now being fostered throughout the new style of video games, where a simple live/die proposition is replaced with experience as the new way of winning.
- Wikis (from the open source movement) represent a departure from the top-down broadcast of information, rather it is bottom-up creation.
- Networking with a selfless attitude of "how can I assist" offers the best way of forming genuine, long lasting business associations and contacts. This is known as "netweaving."
- Ethics is an area linked to your own internal moral compass and that of the organization you're working for.
- Unethical behavior has a disproportionate ability to damage an organization.
- Ethical decisions are not necessarily the choice between right and wrong, but can be between bad and worse.

- Making an ethical decision may take more effort than an unethical one.
- Making an ethical decision tends to be self-reinforcing: the more you do it, the more likely you are to do so in the future.
- Collaboration with non-traditional partners can improve idea exchange and generation.
- Sustainability is about more than just environmentally-friendly solutions; it's about the ability to continue to do things into the future.
- Sustainability has different connotations from the business and public's perspective.
- The product lifecycle embraces the whole life of a product, from its initial conception through design and build to retirement and disposal.
- The learning effect is the reduction in resources required to complete repeat units of the same product, given that each subsequent one brings with it the knowledge to do it better.
- In communication, "context" refers to the extent to which what is said is meant.
- Understanding another's culture can rarely be done from afar; it requires an immersion.

Actions to take away

Some things you can do to act on the information presented here:

1. **Consider** the statement (based on Thomas Friedman's comment to the U.S. National Governors' Association): *"the greatest skills a student can master is to learn how to learn."* Do you agree or disagree with this? Why/Why not?.

2. **Go** to the site www.wikipedia.com and browse some of the listings. Why is this site referred to as "open source"? What does it mean for the site to be "bottom-up" in information content? List three positives and negatives about such an approach.

3. **Ask** yourself: What does it mean for an organization to be sustainable?

4. **Write** down your own definition of what lifelong learning is. Would you say that you are engaged in lifelong learning? Why/Why not?

5. **Describe** what is meant by "context" with respect to verbal communication. With reference to context, how would you consider your communication style, and why?

9. So Why Aren't Engineers...

> There are many questions about the engineering profession that are rarely answered, largely due to their generic, non-technical nature. They usually take the form of "why is it this way and not some other way?" In this chapter, we'll address some areas that commonly fall into this category, namely: remuneration, leadership roles, public understanding of engineering, appearance in the mass media and political aspirations. Understanding why engineers are positioned a certain way can help us understand the possible causes and imagine the ways we can change it.

Paid more?

Salary is usually one of the quickest, easiest and most universal metrics used when comparing professions. Have you ever stopped to wonder why engineers don't get paid more than they do? While engineers are certainly not at the bottom of the workforce when it comes to salary --- not by a long shot --- they are also not as well paid as others. Why is that?

Within a technology-focused organization it would seem logical to award the highest salaries to those people who know, understand and develop the technology --- after all, this is the reason for the organization's existence, right? Well, not really, and the following answer may help give you a clearer understanding as to why engineers are remunerated the way they are.

A technology example I've used throughout this book is the cell phone, a device we use today that is almost synonymous with everyday living. Increasingly, our lives are stored on these devices, and they have shrunk to a size barely bigger than a box of matches. Engineers who understand how a cell phone works functionally and how to design one physically are in demand. Those who understand the technology and how it interacts with the network are in higher demand, but those in the highest demand understand not only the technology itself but also what it is used for. In other words, they know what the customer wants and where these wants might lead them in the future. They are in effect able to project where future demand is most likely to come from. These engineers have both technical acumen and business savvy.

Engineers who understand both the technology and the customer will usually serve on multi-disciplined project teams. You won't find them poring over technical specifications or debating the intricacies of one particular design over another, but they'll know enough to interact intelligently with those who do work these issues day in and day out. Their knowledge will have more of a high level context to it. For example, when looking at specifications, rather than thinking: "How can we make this the best technically superior cell phone," they'll be thinking: "Does what we understand of the customer match the product we are developing," or "Can we extrapolate what we know about the customer to date to meet demands they themselves haven't realised yet?" These are the engineers who tend to receive higher levels of remuneration because they are able to tie together the technology with the business need.

The salary one receives is tied to the perceived value their position can add to an organization. Are your knowledge and abilities used by the company to directly impact the level of revenue they generate, or the competitive advantage they have over their competitors? When we cast an eye across to other professions, such as accountants or lawyers, you can start to understand why they tend to be more prolific in the senior ranks --- they deal with the business angle of the company when first entering the workforce; they see technology as a way to meet customers' needs, not as an end in itself. As engineers, we only start to see the business angle after many years into our careers, or if we go out and actively seek it, open ourselves up to it, and choose to explore more than just the technology.

Leading more organizations?

The previous section concerning what engineers are paid and why points to the reason technically-focused engineers rarely rise to the top of organizations. Think about what it means to be at the top of an organization: you need to be able to lead, have a strategic vision, a degree of personality, good communication skills, a good understanding of the business environment, a good idea of the market the company is in, and finally an understanding of the products or services you are selling.

Of all the attributes desirable in a CEO, there are really none that necessitate a strict knowledge of engineering. They tend more to be business-oriented attributes, which explains why those who go through the organization on the business side tend to rise to the top, while engineers tend to find themselves stuck, for example, at the Chief Technologist level. Understanding technology is one thing, but with nothing more in your repertoire it's not a good idea for the organization to put you into management. Having a technology junkie or zealot at the head of an organization would likely drive the company into bankruptcy, developing a technologically-superior product while oblivious to the fundamental consideration of market need.

Understood better by the community?

Ask five different people what an engineer does and you'll probably get five different answers. You cannot simply tell someone what an engineer does, due in large part to the exceedingly wide reach they have, changing between different organizations and companies and even within their own discipline. People have a good idea what a doctor does because doctors do similar things the world over, and people have a good idea what an accountant does because accountants also do pretty much the same thing the world over. But how do you quantify what an engineer does? One might say they're someone who deals with technology, but how does that differ from a mechanic or technician?

As addressed in the outset of this book, what engineers are and what they do are not necessarily the same thing. Because of this, I don't think it's possible to explain to the community what an engineer is. The best we can do is stop trying to be as rigorous about it as our engineering training directs us to be, and simply say that engineers are people who solve problems --- and leave it at that.

Featured more in mass media (movies, TV shows, etc.)?

I apologize if I'm starting to repeat myself, but the reason more engineers aren't featured in the mass media is due to a lack of understanding of what they do. Movies such as "A Few Good Men," or programs like "ER" feature stars like Tom Cruise or George Clooney because what lawyers and doctors do can be easily defined. We can relate to them because we've all had first hand experience with them. Engineering on the other hand sinks into the background; it is everywhere and therefore, unable to escape its influence, we lose the ability to perceive its existence in a tangible way. We know where to put lawyers --- they go in the courthouse, and doctors go in the hospital, but where do engineers go?

Another reason engineers aren't featured in the media more is that many of them can be very intense. Intense is good, because they are passionate about what they do --- but being passionate about something, trying to communicate it to someone who has no idea what you are talking about, can give the same impression as someone from an insane asylum! If this passion was toned down along with the messages so people could see past the ranting and raving and focus on what's being said, they might stand a better chance of understanding what's being done.

A third reason for engineers' absence from the media is that the work they do tends to be about process, the creative bits in the middle that contribute to an output. Things don't necessarily happen all that quickly in engineering, whereas in "ER" there's excitement and drama that can fit into half-hour episodes. The classic case of the patient having a subdural haematoma (a term that's burned into my subconscious, as every hospital movie I've seen has one) has to be dealt with by the doctor right there and then; there's action, there's movement and there's some kind of resolution all within a short time frame. In the classic lawyer movie, they can cross-examine and pontificate to their heart's content because there's progression, claim and counterclaim, and a resolution which is reached within a finite time frame. It would be hard, however, to imagine an engineering drama: a major design or capital acquisition program that lasts for 26 weeks, each week showing another milestone you've gone through to get to the end of the project, with someone developing Powerpoint slides for their next presentation; "tune in next week and watch Fred compose 30 slides for a five minute presentation, then run way over time and annoy the boss!" This is precisely what we have Dilbert cartoons for: to deride the fact that engineering does not always follow a neat linear path that works to a finite time frame.

A show about engineers? Not quite.

While there is not a full-fledged program about engineers, "Numb3rs" does get us pretty close. From the CBS website's program description:

"...a drama about an FBI agent who recruits his mathematical-genius brother to help the Bureau solve a wide range of challenging crimes in Los Angeles. Inspired by actual cases, the series depicts how the confluence of police work and mathematics provides unexpected revelations and answers to the most perplexing criminal questions. A brilliant mathematician who, since he was little, yearned to impress his big brother. His big brother, Dan, as a seasoned investigator, deals in hard facts and evidence, whereas Charlie, a math professor at a California university, functions in a world of mathematical probability and equations. Now, despite

> their disparate approaches to life, Don and Charlie are able to combine their areas of expertise and solve some killer cases."[1]

In politics?

Engineers involved in politics --- it seems both counter-intuitive and counter-productive. Why would one interested in politics bother to study engineering in the first place? Why not study political science or economics or law, which would seem to be much more appropriate for a life in the political sphere? While it would seem that engineers have no place in the political process, they have more to offer than you may realize.

In the West, politicians tend to be lawyers, self-made businessmen or former heads of big business, but very few have any kind of technical background. There are exceptions, notably the current Chancellor of Germany, Angela Merkel, who has a background in science. It's a different story with governments in the Eastern Hemisphere. In China, all the members of uppermost tier of the communist party are engineers. In India, the current Prime Minister is an engineer. Having engineers involved at the top helps guide decisions made by the leadership that may impact the health and well-being of technology-driven industries or the development of infrastructure.

Engineers are also extremely useful in the legislative process where government regulations are formulated. Having a roomful of politicians with no technical background draft bills that affect the regulation of the power industry would not seem to be the smartest way to do things, and yet this is precisely what has happened in the U.S. As mentioned earlier, some engineers have made a very lucrative career translating technical documents and arguments into language politicians can understand and incorporate.

Engineers generally stay out of politics and the political process because it's not so cut and dry; there's no right and wrong, and there is a lot of negotiation involved. Many engineering arguments need to be justified in terms of their cost or economic impact rather than their technical merit. The political process and politics in general can also be a particularly fickle and unforgiving occupation where perception counts far more that facts.

Chapter 9, if you only have 5 minutes...

- Engineers are often not the most highly remunerated professionals in an organization, even technology-based ones.

- Engineers possessing technical acumen along with an understanding of the business are highly prized, as they are able to connect an organization's technology offerings to customer wants and needs.

- Within an organization, salary is expressly connected to the perceived value of that position.

- Accountants and lawyers tend to be represented in senior ranks because dealing with the business aspect of an organization tends to be their primary focus from the early part of their career, while for many engineers dealing with this aspect may not happen for many years.

- Strict knowledge of engineering is not typically a requirement to be a CEO, but knowledge of business is.

- Technologically superior products are useless unless they address a market need.

- Public understanding of engineering is typically low due to the lack of an easily understandable and commonly held function of engineers.

- Engineers are professionals who solve problems.

- Engineers are not featured much in the mass media due to the difficulty in being able to define what they do.

- As engineering pervades everyday life, it is difficult to observe its effects objectively from a vantage point outside its influence.

- Engineers' passion for their subject matter can sometimes obscure their message when trying to communicate with non-engineers.

- Engineers typically shy away from politics because there are no objective truths, merely opinion and perception.

Actions to take away

Some things you can do to act on the information presented here:

1. **Go** to any salary site like salary.com and input a particular discipline or job title, e.g. electrical engineer.
2. **Think** about whether the pay scale reflects what you think the position should pay. What can you say about the pay scale and the associated job description?
3. **List** the names of three engineers you can remember from TV shows or movies.
4. **Identify** the traits and characteristics demonstrated by these characters. Are these attributes what you would expect to find in an engineer? Why/Why not?

10. The Successful Engineer's Secret—Attitude

> Your attitude will impact your success as an engineer just as much as the tangible skills you bring to the table. In this chapter we'll consider how attitude can influence aspects such as goal setting, time horizon, career direction, self-confidence, perception of success and failure, physical well-being and sense of purpose.

Goal Setting

The goals you set for yourself today will likely be different than the goals you set for yourself 20 years from now, due to the influence of unforeseen and unknowable future events. For this reason you will need to periodically take stock of your situation and question whether your goals are still valid.

At these times of reflection, it's often beneficial to gain the perspectives and thoughts of a close friend or colleague. This process of obtaining an outsider's perspective can be cathartic. The very act of talking about your future goals and concerns can reduce the anxiety associated with them. Be prepared, however, for critical judgment from those you are seeking assistance from --- they may tell you that your situation, and ultimately you, are somehow outside of the norm.

Most people seeking advice just want the listener to provide alternative ways of thinking so they can gain a better understanding of their problem. Due to feelings of vulnerability, problems can arise when responses such as "That's a silly idea," or "Why in the world would you want to do that?" are received. These types of responses can create hesitation in the person asking for advice, and they'll be less likely to ask for help in the future, either from that person or anyone else.

Therefore, before progressing any further, let me be clear: it's not my job to judge you. Regardless of your situation, the fact that you're reading this book means you want to do something about your situation --- you're one of the VERY few who act, rather than merely talk.

Perception of time

I remember being a teenager and thinking about what it was going to be like in my twenties. People in their twenties seemed so old, and there was no way that I wanted that. I thought by the time I hit my mid-twenties I'd definitely be old. Everything seemed to speed up after graduating from high school, and the years zipped by at an ever increasing rate.

Similarly, when people talk about a "career," they typically think in terms of long time frames --- but when you stop to think about it, your career doesn't last very long at all. Let's assume you start working at 25, after you finish university. Men and women usually formally retire between ages 60 and 65, depending on culture and country. This means you have about 40 to 45 years in the workforce. (This figure is being extended as many people are working longer now out of choice or necessity, but we'll stick to the conservative estimate of 40 years.)

Just as the passage from teenager to twenty-something comes before you know it, so too a career of 40 years is going to go by pretty fast. The time from when I finished high school to now seems to have flashed by --- and it's been 17 years! There's no point in denying the incessant march of time; all we can do is make sure we use time the best we can. This means realizing that your career --- what you're doing now --- is a one shot deal. Since this is the way it is, then why not give it everything you've got?

Direction

So, which direction do you want to go? Many just starting out have little idea what they want to do and while lack of direction is not good, neither is being unwilling to consider directional shifts. As mentioned earlier, constant bombardment from outside influences (media, parents, school teachers, friends and so forth) unconsciously influence the opinions you form and the directions you set. Your personal circumstances may also have an impact; moving to another country, getting married, having children, being in an accident and winning awards or scholarships to list but a few examples. When either personal circumstances or other influences channel your career in new directions, it becomes a matter of determining your future path from the new set of co-ordinates.

In determining your direction, you'll undoubtedly receive lots of advice from people concerning what your career path should be, where it should go, and what the best jobs are for you. In the end, you're the one that's got to live with the decision --- you're really the only one who knows what's right for you. It's best to be receptive to people's ideas, thank them for the suggestions and then make your own decision whether or not to follow their advice. In this way, the suggester feels like they've offered some help, and you've graciously accepted their advice without telling them you don't like it or won't be using it. Taking this approach becomes more difficult the closer the person is to you.

Unfortunately, many students enter university, jobs or other endeavors as a means to please their parents --- ironically, this rarely works. In pursuing an endeavor that isn't yours, you'll never really enjoy what you're doing and by feeling trapped in the pursuit of pleasing your parents, your innovation and creativity will be stifled and it will become a chore to continue. Some parents seek to live their lives vicariously through their child by pushing them into professions of high status as a way of getting to a goal or career they were never able to realize. While many parents aspire for their sons and daughters to be doctors, lawyers or dentists, I believe what they really want is for their kids to engage in a profession that is well respected, that they enjoy, and will reflect well on the job they did as parents.

Success

Everyone has a different definition of success. While popular media may focus incessantly on the external world, real success comes from internal contentment with your situation. Looking to the outside world to validate your views of success can sometimes "lead you down the garden path," and in so doing you may find yourself hostage to someone else's ideals.

Competition

As director/actor Woody Allen once said: "Eighty percent of success is just showing up."[1] While our primary concern may be about beating another person to the finish line, we're ultimately only in competition with ourselves --- so just being prepared to give something a go will usually mean that you've given yourself a better than even chance of doing well.

Journey versus the destination

Is "success" defined as reaching the goal, the endpoint --- or is it the journey travelled to get there, the process? Let's assume for a moment that success as an engineer is reached by attaining the goal or endpoint. If this was the case we'd always be looking ahead for that next patent, degree or promotion, all the while detesting the time it took to get us there. We'd have little attention on what was happening to us today. This situation can best be described as "living for later" versus "living in the now."

> **Now versus later**
>
> The predicament of "living for later" versus "living for now" became the focus of the 2006 comedy film *Click* starring Adam Sandler. Through the use of a magical remote control, the main character fast-forwarded his everyday life to arrive at the achievements he thought he wanted. Before long, he realized his life was made up not of his accomplishments, but of the roads taken to get there and that this was where the true success of his life was found.

Coaches often talk about runners who "run on guts," or say their sport is 10 percent physical and 90 percent mental. While it is necessary to have some base level of ability to be an engineer (or an athlete), it's not just ability that determines how far you'll go --- it's whether you enjoy the process of "being" and the challenge it presents.

Research has disproved the notion of being born with a natural gift or talent. Extensive research conducted in the U.K. on the linkage between talent and other factors, as reported in *Fortune* by Geoffrey Colvin, concluded: "The evidence we have surveyed... does not support the [notion that] excelling is a consequence of possessing innate gifts."[2] In essence, researchers found that those who were great at something became that way because, unsurprisingly, they worked the hardest.

Consider Tiger Woods (putting aside his recent personal struggles). He's a champion golfer because he practices more than anyone else, even to the point of being unreasonable. The success he has with golf today was gained by spending hours and hours working on every aspect of his game. And yet, if he didn't enjoy the "process" it's doubtful he would expend all that time and energy just for those fleeting moments when he's actually in the process of winning. When competing, he spends most of his time "playing" the tournament; it's only at the end, when he's been announced the winner, that he actually experiences "winning."

Lance Armstrong is a more extreme example. Lance has won the Tour de France seven times, more than any other person. Yet there is no way Lance would have gone through such as a gruelling training regime and strict diet if he didn't actually want to experience going through it. The journey, initially thought of as secondary, becomes the focus and the end result (winning) almost becomes a by-product of that journey. Lance Armstrong, Tiger Woods and others like them have been tagged by Thomas Friedman, author of *The World is Flat,* as "strategic" winners,[3] that is, those who look to win for the long term.

How does this relate to your career as an engineer? Well, you will certainly have moments basking in the glory of "winning"; receiving an award, being top of the class and receiving recognition from your work colleagues or a professional society for example. As nice as these moments are, you must truly love "being" an engineer, and constantly seek to improve yourself at every turn (like Lance Armstrong or Tiger Woods) in order to become a strategic winner, not just someone who wins at a particular point in time and then moves on when the glow has faded. An engineer's success lies in "being" an engineer, doing your job, solving problems to the best of your ability --- external success will naturally come as a by-product of this, and all the while you'll be the happier for it.

Trust your own counsel

When looking back on my "early career," I have come to notice how my opinion changed depending on the phase of life I was in. As an example, look at how your opinion of your parents changed as you transitioned from childhood to teenager to adulthood. When you're a kid your parents' opinions are 100 percent fact, they're gospel --- there was no thought in your mind that things your parents told you were colored from their own perspective of the world and could be wrong. As we get to adolescence and progress through our teenage years our parents' words lose their sacredness, and in some cases we may go against everything they say. We don't believe a word they utter, and whatever their opinion is it's just plain wrong.

An analogous experience to the relationship with our parents is faced by many at university. As an undergraduate student you tend to take everything told to you by lecturers as gospel but this changes and as you enter the workforce, you realize that the information they gave you was colored from their own perspective of the world. The message here is that no matter how smart or clever people appear to be, they are subject to the same vulnerabilities as all of us --- the information they provide is not the absolute truth, but their version of it. This means your opinion has just as much validity and value as anyone else's.

More than once I've taken the counsel of someone I thought to be knowledgeable about a particular subject only to realize later on that they really had no better handle of the situation than I had (and in some cases less). The inferiority I felt about my view on the issue was completely unfounded. I really just needed to trust my own judgement and not worry about what other people might say, think or do.

It can be hard to have self-confidence in your views as a young engineer, particularly when faced with alternate opinions from those who are older and appear more experienced. I have often found, however, that the more strongly a belief is held the less likely it is to be right, or the less likely alternate points of view have been considered.

I'm sure at some point in your life your parents, career counsellor or other person you sought advice from said you must have a plan, have thought things through, and have a backup plan in case all else fails. In general this is sound advice as it ensures you have a safety net. However, there are also times when you've just got to trust yourself and barge ahead believing you will succeed no matter what. Doing this is a scary proposition, and it's not something that's going to work all of the time.

> **Gattacca**
>
> An example that encapsulates such self-assuredness comes from the 1990s movie "Gattacca," starring Ethan Hawke and Uma Thurman. In one scene, Ethan Hawke's character Vincent competes against his brother at the beach in a swimming race. They'd done this many times before and his brother always won because he was genetically pre-disposed to win. This time, however, Vincent's determination, courage and belief in himself allowed him to swim further than his brother (who ends up being saved from drowning by Vincent). Just before the brother succumbed to fatigue and sank beneath the waves, he yelled: "How are you doing this Vincent, how have you done any of this?" He could not fathom how Vincent achieved his long term goal of becoming an astronaut with his 98 percent probability of a heart condition. To this Vincent replied: "You want to know how I did this? Here's how I did it: I never saved anything for the swim back!" He couldn't worry about the game plan or what his brother might do, he could only control his own actions. He was so confident in his abilities and of attaining his goal that he went from being reasonable to unreasonable about its achievement. He pursued his goal past the point that any reasonable person would pursue it, but it was only when he got into that unreasonable state that it became a reality.

This is the essence of what I'm trying to say here: You've tried reaching a goal before and failed... So what? That goal is still out there and you still have a number of different ways to achieve it. The only limiting factor to you achieving it is you, and while this may seem unreasonable to other people, even almost ridiculously so, they may just be surprised when you pursue it to the point where, in that zone of unreasonableness, you actually achieve what you set out to do.

To further illustrate my point, I'll again borrow a scene from a movie, this time the 1980s Oliver Stone film "Wall Street." This film highlights what persistence can get you. Gordon Gecko (played by Michael Douglas), the main character in the film and hero of young up-and-coming stockbroker Bud Fox (played by Charlie Sheen) tries to relay to Bud what it takes to succeed in the business world: "Most of these Harvard MBA types don't add up to dog shit. Give me someone who's down on their luck any day of the week." When you are wedged under the heel of poverty or personal ruin, things start to come into focus real quick. You've got two choices: stand up and see your circumstance as a challenge to be overcome, or lay down and let your circumstances steamroll over you. If you choose the former, you start to get real creative and your levels of perseverance and determination shoot way up because you realize that this is it: "I may not get a second chance and my existence depends on it."

The source of these feelings tends to extend from the previously mentioned Maslow's hierarchy of needs. The closer you get to the base levels of this hierarchy the greater the desperation and lengths people will go to in order to satisfy it. Using a personal example from my past, when I first came to the US, I was starting to get pretty creative after five months looking for a job and not getting any responses. I decided to attend a space conference in Long Beach, California to see if I could meet some people and make some inroads.

One panel discussion at the conference involved the CEOs of Space Systems Loral, Orbital Sciences Corporation, Lockheed Martin Space and Missile Systems, Raytheon and Ball Aerospace. As they delivered their take on the state of the industry, I decided to put a question to them that would hopefully elicit a favorable response. As it happened, mine was the last question of the session and the moderator addressed it to the whole panel. It read: "I am an Australian with an engineering undergraduate degree, a technical masters, an MBA, have served in the Australian Air Force for 5 years and have had a SECRET clearance with the Australian Department of Defense. I have been in the U.S. for 5 months and cannot get an interview. Why not?" The gentlemen from Raytheon, Lockheed Martin and Ball remained tight lipped, which did not bother me since I knew that every contract they had was with the U.S. government and therefore required U.S. citizenship. The CEO of Space Systems Loral spoke into the microphone: "There's definitely room for someone like that in our company. Please come and see me after the session." Next, the CEO from Orbital Sciences leaned forward: "You haven't got an interview because you haven't called me yet!" He then gave out his office number and asked the questioner to see him afterwards. After the session had finished, I introduced myself to both the CEOs and both promised they would be in touch in the coming week.

Within two days I was approached by a director from the Johns Hopkins University Applied Physics Laboratory (APL) who told me about several opportunities there, and said I should consider working for them. In addition to this several other doors opened up, and it was all because I took an unconventional approach precipitated as a direct result of my dire circumstances.

What my experience demonstrated is that in many cases, your audience simply doesn't know you're there. For whatever reason you're just not being heard, and you need to do something to turn the volume up and make your presence known. Although I'd been applying for positions left and right, it was in a very bottom-up fashion via the internet so my name was only getting to a few companies. Asking that question at that conference put my case out there in front of 1500 people. Every person I talked to after that session asked if it was my question, since I was one of only a few Australians in attendance.

Although the situation at the conference turned out positive for me, my question could just as easily have not been read and sometimes you're going to fall and fall hard no matter what you do. When this happens, there's nothing you can do but take it on the chin and realize that the only failure is not getting back up. An example that exemplifies this is the case of Carly Fiorina, the former CEO of Hewlett Packard who was unceremoniously dumped in 2005. Since then, she has gone on to write a book, "Tough Choices", was an advisor to 2008 presidential candidate John McCain and at the time of writing, is contending for a seat in the US senate. When asked about being dropped as the CEO from HP, she said: "The worst thing I could have imagined happened. I lost my job in the most public way possible and the press had a field day with it all over the world. And guess what? I'm still here. I am at peace and my soul is intact."[4] So you see, for all their polish, failure even visits the CEOs of large companies.

Turning failure into success

There will be times in your career when you know you have not done a job very well. For whatever reason, you haven't delivered as well as you thought you had and you don't know how to make it better. Someone stepped in after you and cleaned it up, made a great success out of it, and made it look effortless. Now you're doubly depressed: not only did you do a bad job, but someone else's success reaffirmed how badly you did.

Such a situation has happened to me more than once. I vividly remember the time I was chair of a young engineers' committee. I wasn't achieving a great deal, partly because the committee members weren't enthusiastic about taking the lead on things and partly because I was to hand the role over to someone else in the coming months.

When the new chair took over, he instigated some great initiatives, far more than I thought possible. This caused me to reflect on my performance as chair: "If he's successful at it and I wasn't, what does that say about me?" Because someone had done a better job than I had, I assumed I had failed. However, I hadn't realized that I was spread too thin; in addition to my role as chair, I was also balancing other committee positions and multiple other endeavors.

The physical—healthy body, healthy mind

I remember sitting on the couch one afternoon with a bag of heavily-salted barbecue chips watching *The Oprah Winfrey Show*. She was speaking with Arnold Schwarzenegger. During the interview he said something extremely insightful that forced me to sit up and stop shovelling chips into my mouth for a few seconds, something I wasn't expecting from *The Terminator*. He said something like: "If you're not exercising your body you should be exercising your mind, and to do neither is to waste time." That encapsulated a logical and easy path to follow. Sure, there will be downtimes (like eating barbecue chips and being a slob on the couch) but in general, if you try to keep body and mind in focus you'll rarely go wrong.

School counsellors, parents, and numerous men's and women's fitness magazines bombard us with the virtues of leading a physically active life, but beyond the anecdotal evidence there is scientific proof of a connection between physical activity and the ability to carry out complex tasks of the kinds engineers regularly undertake.

Research conducted at the Biomedical Imaging Center of the University of Illinois proved there was a "constant, reliable effect" from exercise on cognition.[5] Utilizing brain scans, the group showed that "increases in cardiovascular fitness in humans increased the level of activity in the part of the brain associated with successful task completion."[6] The study continued: "Aerobic exercise is most likely of greater value, researchers say, possibly because it increases blood flow and oxygen to the brain."[7] The proof is in the research, but there are also examples from everyday life.

In the November 2005 edition of *Runner's World*, a story titled *Running the Company* profiled the CEOs of some of the world's most successful companies; all were avid runners. When asked about the synergy between runners and CEOs, Dennis Carey, a partner at an executive recruiting firm, commented "It's about integrity, hard work, passion, hands-on dedication. All the things that if you had a position spec for a great runner, it would be the same for a great CEO."[8] Indeed, running to reach a goal such as completing a 10k or marathon teaches the runner about perseverance, commitment and humility, and that nothing comes easy. "Successful people think of long term horizons, whereas the unsuccessful think more about the immediate. Anticipating the future lets them get through the pain of the present."[9] Having long term horizons and looking beyond short term losses is another approach that tends to "level people out." They tend to be more mellow when they're not upset by day to day affairs, but instead have their eye on the bigger picture.

So if activities like running enhance your abilities as an engineer, are there any that can make you less effective? Well, you'd probably say drugs or alcohol or similar destructive behaviors, and you'd be right; but something far more insidious, that you've probably had intimate experience with while at university, is a lack of sleep.

When you're at university, particularly studying engineering courses, sleep becomes a commodity you think you can trade for more study time, but unfortunately the extra time you spend studying yields little to no return. In an article published on ScienceCareers.org, author Irene Levine explored the effect lack of sleep had on various people. "I'd rate lack of consistent sleep as the single most consistent detractor of my productivity," said one assistant professor at a large research university[10].

I can attest to the negative effect lack of sleep has on output. I was trying to finish my thesis during the final year of my engineering degree. It was common practice for final year students to pull multiple all night sessions before completing their thesis, and despite repeated calls from our lecturers not to fall into the same trap, my fellow engineering colleagues and I all did. I stayed awake from Tuesday morning until Thursday lunch time (about 50 hours) before handing in my thesis. I distinctly recall the phases I traversed in this sleepless ordeal. First there was determination, propped up by the rush of doing something that seemed indicative of strength (although clearly it highlighted how I'd underestimated the workload required to finish). Then I started to notice little things becoming magnified --- for example, the whirring of the cooling fan in my computer became almost deafening, and my actions felt heavy and strained like the puppets on the old Thunderbirds TV show. I went through periods of light-headedness, and of course my eyelids felt so, so heavy. I also had to concentrate to perform even the most routine of tasks, like reaching to grab a pen.

As I headed into the final stretch, the adrenalin kicked in and I felt more alert and awake. After handing in my thesis I remember thinking that I wasn't so tired after all, and that although I'd go home and get some rest I could probably stay awake a little longer. That's when I realized how tired I really was. I went home to bed and fell into the deepest 14 hour sleep I ever had. My parents came around to my flat the next day to pay me a visit and despite persistent banging on my door could not wake me. Thinking that something was wrong, they asked my next door neighbor to climb through my bathroom window and check to see if I was okay. I woke up to see this man in a handstand position as he came in head first through the window. I was completely disoriented, had no idea what day it was or where I was. So if there's one lesson to be learned from this, take it from me --- skipping on sleep is definitely NOT an option, because it will just end up detracting from your output and performance. Also, make sure you start your final year engineering thesis early!

Balance

At one of the first professional society career fairs I attended[11] upon coming to the US, a hiring manager relayed an anecdote she once heard from a senior executive of The Coca-Cola Company that perfectly described the struggle we all face in finding a work/life balance; imagine you are juggling five balls, trying to keep them all in the air at the same time. The five balls represent work, family, friends, health and relationships. The work ball is made of rubber such that if you drop it, it always seems to bounce back. The other balls, however, are made of glass --- if you drop them, they get scuffed, chipped, and perhaps irrevocably broken.

This is the best analogy I've heard about the work/life balance. We tend to focus so much on our work, not realizing there is a degree of elasticity in it, while paying too little attention to our friends, family, health or relationships, areas which may have very little elasticity --- and if damaged, we may not be able to recover. I'm sure you know people who bury their heads in their work to the total exclusion of these other areas of their lives. While they may find a degree of success in their job, the victory is turned hollow by the complete neglect of these other areas --- they have no friends, their family is distant, their health is in disarray, etc. In a paradoxical way, spending time ensuring these other areas are in order can give you greater zest for work -- it can even make you more successful than constantly keeping your nose to the grindstone.

Dealing with disappointment

If you view yourself (or others view you) as a high achiever, then you're accustomed to chasing after goals and achieving them. Most articles on the subject will tell you to reach for the stars and not to lower your sights, a position I also wholeheartedly advocate. In today's world, the prevailing thought is that if your workplace doesn't offer you advancement or satisfaction, then you should look for some place that can.

This is an okay approach to take if you're new to the organization, but what if you've already 'moved on' three or four times and are still not satisfied with your work environment? Many bosses are going to see a resume with a number of 6 to 12-month stints on it as an inability or unwillingness to commit --- particularly if your next job is with a larger, more conservative company. So while there should be no question about staying in a job you detest, there may come a time when twenty- and thirty-somethings may need to tough it out in a particular position rather than jumping at the first sign of trouble.

My own experience reflects this. I resigned my commission in the Air Force after five years as an officer because my career was not going where I wanted. I went out on my own as a consultant for two years before jumping on board with another consultancy, sticking with them for another year before leaving for the United States. While I left these positions with forethought and assurance that it was what I wanted to do, my resume tended to give the wrong impression. For this reason I believe our drive to succeed can sometimes be our Achilles heel. Our unwillingness to settle down or be complacent for even a short period of time can be our downfall. Instead of allowing events to unfold naturally, we try to force through them with our sheer strength of will.

Reading

I remember a Biology teacher I had in my first year of high school imparting some advice to my parents that I have put to great use in the almost 20 years since: Read. He advised my parents to encourage me to read whatever I could get my hands on, no matter what it was. In our modern age of bite-sized information, the good old-fashioned book seems to lose out to online content far too often.

The type of reading I'm talking about is not the "skim" reading you do because you have an assignment due and you have to read this or that article. I'm talking about silent, focused, deep, contemplative and immersive reading, just you and your thoughts. And you don't want to just read things you agree with --- read those articles whose point of view you oppose. Doing this allows you to consider many viewpoints besides your own and this is extremely valuable in engineering. While we like to think there is only one correct science-based answer, engineering is a human endeavor and when you work with others, they are going to see things differently than you. If you are unable to put yourself in their shoes and bridge that gap then you're never going to be able to work with others in order to reach a common objective or goal. Reading is also good downtime. The pace of life is so frantic that the only way such a pace can be maintained is for it to be interspersed with periods of downtime to allow your brain to rest and recharge. Reading should therefore be a leisurely activity rather than one accompanied by speed; take time to absorb what you are reading and let your imagination work on it.

Finally, reading broadly gives you a much richer communicative ability. You are able to draw on a greater reservoir of examples and supporting arguments when communicating ideas to others, something engineers can be less than proficient at if all we focus on is engineering.

Single mindedness

There's nothing wrong with a serious sense of purpose in relation to something you want to achieve. Those goals that are presently out of reach or harder to obtain are going to require focus and perseverance if you stand a chance of achieving them. Long term goals may take months or years to attain, and may involve numerous roadblocks and setbacks which prevent you from achieving them. Engineering is an intellectually taxing endeavor, and it's natural that you will find it difficult at first -- it's meant to be. When you think it's all too much to bear, just realize that you are one of the very few who have the privilege to study engineering. There are many others in this world who never see the inside of a classroom and will never know what it is to receive an education.

Chapter 10, if you only have 5 minutes...

- It's natural to feel a little lost and confused regarding your professional direction in the first few years in the workforce.
- Goals change over time, so it is constructive to review them periodically to determine their relevance.
- Discussing professional concerns with a colleague is a good way to obtain some independent and/or outsider's perspective from someone who is familiar with your situation.
- Discussing professional concerns helps reduce the anxiety that may be associated with them.
- Sometimes discussing an issue with a colleague can represent a means of understanding your own "problem space."
- Attitude is a large component of how you deal with problems in your career.
- It is unreasonable to expect that you will have all the answers figured out in the first few years of your career. Being an engineer is a process that unfolds across many years.
- Despite the perception upon beginning your career that you have plenty of time, your career will move pretty quickly.
- It's important to remain flexible in your career as events outside your control will inevitably impact you.
- Which direction your career heads is entirely controlled by you.
- Rather than outwardly rejecting people's advice, it's far better to receive it and then decide whether to act on it.
- Make sure that being an engineer is what you want to do, not something you're doing to please another person.
- What constitutes success in your career can only be defined by you.
- Your career will be about the journey travelled rather than the ultimate end result or destination.

- Natural talent or ability is not the largest determinant of your success in a particular endeavor --- working at it is.
- Success as an engineer comes as a natural result of doing the best you can regardless of the situation.
- It's important not to assume that someone else's judgement (no matter how senior or experienced they may be) is better or more informed than your own.
- While having a plan is often necessary, sometimes a leap of faith or trusting your own abilities to meet the challenge, no matter what, is required.
- University research has shown concretely that aerobic exercise has a positive effect on cognition.
- Keeping an eye on the bigger picture will ensure that you do not get too upset by day to day troubles.
- A large contributor to daily success is getting a regular, uninterrupted eight hours of sleep per night.
- Maintaining a healthy balance across work, family, friends, health and relationships can help optimize a feeling of centeredness.
- A resume replete with short stints (6- to 12-months) at different organizations may be looked upon as an unwillingness to commit, and may be viewed negatively when looking for a new position.
- Reading widely, critically and thoughtfully is an excellent way to expand your horizons in multiple dimensions.
- Long term goals/endeavors may take many years to attain and be replete with numerous roadblocks.

Actions to take away

Some things you can do to act on the information presented here:

1. **Take** 3 minutes and write down 5 goals you would like to achieve professionally.

2. **List** 3 goals that you've attempted in the past and not yet achieved. Try to define what it was that prevented you from achieving it, or what the catalyst was for falling short.

3. **Decide** on a goal (either of the 3 listed or another beyond that) that you've attempted in the past but have yet to achieve which you still want to accomplish.

4. **Ask** yourself openly and honestly: how physically active are you? Do you feel this is something you've neglected to this point? Why/Why not?

5. **Record** over the next 5 nights how much sleep you get per night. If you don't get 7 to 8 hours sleep per night on a consistent basis, you need to take action.

Afterword

Much has happened since the first edition of this book was published. In the intervening 7 years or so, a new US president has been elected, social media has gained prominence and become more pervasive in our lives, terrorism has permeated the international landscape, cybersecurity has become an area of concern for many organizations (both public and private) and the world has largely recovered from the Global Financial Crisis, though huge swaths of society are still not without challenge.

Throughout the book, I referenced specific people whom I held up as exemplifying particular desirable traits or behaviors. There are two individuals I feel I should revisit in particular; Lance Armstrong and Arnold Schwarzenegger. Despite years of vehement denials, Lance Armstrong was found to have undertaken a long term, systematic program of cheating and as a result, was stripped of his 7 Tour de France wins. Arnold Schwarzenegger was found to have exhibited his own personal failings (an affair) resulting in the dissolution of his almost 25 year marriage. While it may be easy to point to the actions of both as examples of how their legitimacy as role models has been nullified, my view is that the sentiment contained in their examples is bigger than the people themselves. In other words, the ideas they espouse go beyond and are larger than the individuals themselves, whether they live up to them or not.

Another issue I feel a need to address has been a rise in popularity behind the idea of quitting college and striking out on your own. While this may be more germane to the American experience, the increasing costs of attending a tertiary institution have further served to place the value of a university degree under greater scrutiny. My personal belief is that most of those singing the praises loudest of not attending college seem to do so only after they have graduated and accrued their success (usually financial). It's easy to cast a critical eye on the experience in hindsight but my belief is that attending these institutions are not so cut and dry as to be just a place to gain tangible job-ready skills. It's also about appreciating different points of view, interacting with people from different nationalities, backgrounds and social standings and learning how to apply effort in the pursuance of a goal; graduation. There are of course many examples of people who were successful without college; Mark Zuckerberg, Steve Jobs and the like but are they 'outliers' or representative of the path that people can (and should) expect if they turn down college in preference to following their dreams? In either case, I think they're examples of just different paths taken; neither one should be viewed as better or worse rather just different.

In summary, I believe that the main elements outlined in the book are still relevant or at the very least, represent a sound starting point for further exploration.

Reece Lumsden
Feb 2018

Acknowledgements

This book has seen me liaise with numerous people from many different industries, disciplines and countries and while it would take far too long to list everyone who has provided assistance, as a sign of my gratitude, I would like to extend my thanks to the following:

Alan Hodges	Lisa Lazareck
Andrew Bainbridge-Smith	Mark Montrose
Antonette Joseph	Maryam Al Thani
Bevan Roberts	Michael Chen
Borre Anderson	Mike Evans
Bryan Dansberry	Paul Bonato
Claude Rousseau	Paul Reilly
Darrel Chong	Peter Dickinson
Debbie Pearson	Raymond Findlay
Dora Musielak	Rebecca Barker
Ed Perkins	Richard Tuggle
Edgar Bradley	Sajid Aziz
Gerald Anleitner	Shreekanth Mandayam
Ghislain Dewalle	Sol Dovac
Greg Walters	Terje Egeberg
Harold Frey	Trevor Harding
Jim Wilkinson	Vijay Arora
Kamal Aryal	Vivek Wadhwa
Kristi Brooks	Wanda Reder
Lars Svensson	Zhenry Huang
Liam Waldron	

In addition, I would like to thank Andrew Seltser for helping to review this book.

References

Chapter 1

1. Macquarie dictionary online, Search term: 'engineer', www.macquariedictionary.com.au
2. Macquarie dictionary online, Search term: 'engineering', www.macquariedictionary.com.au
3. Caines, M. *We need a major rethinking not just of what we do but how we think*, Engineers Australia, Vol 75 No 7, July 2003
4. Nasby, G. *So you wanna be an engineer?*, http://www.grahamnasby.com/misc/engineering_about.shtml, accessed 23 November 2005
5. Wadwha, V. *About that engineering gap...*, Business Week Online, http://www.businessweek.com/smallbiz/content/dec2005/sb20051212_623922.htm, 13 December 2005
6. Farrell, D., Grant, A. J. *China's looming talent shortage*, McKinsey Quarterly, 2005 Number 4, http://mckinseyquarterly.com/article_page.aspx?ar=1685&L2=18&L3=31
7. SSTL Press release, *First satellite launched for Algeria in Surrey's Disaster Monitoring Constellation*, http://www.spaceref.com/news/viewpr.html?pid=9948, 30 November 2002
8. Claxton, G. *Wise Up—The Challenge of Lifelong Learning*, Bloomsbury, 1999, pp 256-257
9. A point raised by my year 11 English Literature teacher (Mrs Szota) back in 1991

Chapter 2

1. Samuelson, R. J. *Prestige Panic*, Newsweek, http://www.newsweek.com/2006/08/21/prestige-panic.html, 21 August 2006
2. World Science and Engineering University Portal, http://www.universityportal.net/2007/09/world-university-ranking-of-engineering.html
3. Krueger, A. & Dale, S. B. *Estimating the Payoff to Attending a More Selective College: An Application of Selection on Observables and Unobservables*, Quarterly Journal of Economics, vol. 117, no. 4, November 2002, pp 1492, http://www.irs.princeton.edu/pubs/pdfs/409.pdf
4. Samuelson, *loc. cit.*
5. Usher, A. & Savino, M. *A World of Difference: A Global Survey of University League Tables*, January 2006, pp 3, http://www.educationalpolicy.org/pdf/World-of-Difference-200602162.pdf
6. Illing, D. *Ranking systems found wanting*, The Australian, 22 Feb 06, pp 25
7. Dodge, J. *Engineering Education: The Nature of the Crisis*, http://www.designnews.com/article/6045-Engineering_Education_The_Nature_of_the_Crisis.php, Design News, Issue 4, 17 March 2008
8. Wankat, P. & Oreovicz, F. *A more evenhanded approach to tenure*, ASEE Prism Online, April 2001, http://www.prism-magazine.org/april01/teaching.cfm
9. Dodge, *ibid.*
10. Wilkinson, J. *Tests are no great gauge of learning*, The Australian, 12 July 2006
11. Stossel, J. *Big Cheats on Campus*, http://abcnews.go.com/2020/story?id=264646&page=1, 19 November 2004
12. Erwin, G. *Researching academic dishonesty*, http://www.kettering.edu/visitors/storydetail.jsp?storynum=67, 15 April 2004
13. E-mail Prof T. Harding, 21 November 2005

14. McCabe, D. et al *Academic Dishonesty in Graduate Business Programs: Prevalence, Causes, and Proposed Action*, Academy of Management, Learning and Education, Vol 5, No 3, September 2006, pp 294-305, http://faculty.mwsu.edu/psychology/dave.carlston/Writing%20in%20Psychology/Academic%20Dishonesty/Grop%204/business2.pdf
15. Impactlab, *Study: Business Students More Likely to Cheat*, 21 September 2006, http://www.impactlab.net/2006/09/21/study-business-students-more-likely-to-cheat/
16. Alexander, P. *The internet at school: A tool or a crutch?*, www.msnbc.msn.com/id/11727321/print/1/displaymode/1098, 8 March 2006
17. Loi, L., Yuan, W. *Open Book Examinations*, Nanyang Technological University Singapore, http://www3.ntu.edu.sg/nbs/sabre/working_papers/10-98.pdf
18. Dillon, W. *Study examines why students cheat*, USA Today, http://www.usatoday.com/tech/science/2006-06-26-cheating-study_x.htm, 26 June 2006
19. Arenson, K. *Students receive fewer A's, and Princeton Calls it progress*, 20 September 2005, http://www.nytimes.com/2005/09/20/nyregion/20grades.html?ex=1284868800&en=1f956c4bc36c2b5a&ei=5090&partner=rssuserland&emc=rss
20. Stephen L. Carter, *Civility: Manners, Morals, and the Etiquette of Democracy*, New York: Basic Books, 1998, pp 217
21. Prof B Manhire's homepage, *Grade Inflation*, http://www.ent.ohiou.edu/~manhire/grade/grades.html, 7 October 2006
22. Killeya, M. *Eleven myths about PhDs debunked*, http://www.newscientist.com/article/mg18925392.200-eleven-myths-about-phds-debunked.html, New Scientist, Issue 2539, 18 February 2006
23. Andrews, N. & D'Andrea Tyson, L. *The Upwardly Global MBA*, ISB Insight, June 2005, pp 14, http://www.isb.edu/isbinsight/linsight_June05.pdf
24. Amble, B. *Employers give thumbs down to academic qualifications*, Management Issues, 20 July 2006, http://www.management-issues.com/display_page.asp?section=research&id=3420
25. E-mail from Ms D. Pearson dated 23 April 2006
26. *The candidates answer questions*, The Institute, 1 September 2006, http://www.theinstitute.ieee.org/portal/site/tionline/menuitem.130a3558587d56e8fb2275875bac26c8/index.jsp?&pName=institute_level1_article&TheCat=2201&article=tionline/legacy/inst2006/sep06/fdebate.xml&

Chapter 3

1. Pencinger, C. *Engineers can do it!*, Letters to the Editor, The Institute, March 2004, pp 15
2. The Personality Page, www.personalitypage.com, accessed 15 August 2006
3. Gereffi, G. et al, *Getting the Numbers Right: International Engineering Education in the United States, China, and India*. Journal of Engineering Education, Vol. 97, No. 1, pp. 13-25, 2008. Available at http://ssrn.com/abstract=1081923, pp 14
4. *ibid.*, pp 20
5. Etter, L. *Is General Motors Unraveling?*, The Wall Street Journal, 8-9 April, pp A7
6. Amble, B. *US employees paying the price for benefits*, Management Issues News, 13 October 2006, http://www.management-issues.com/2007/8/31/research/us-employees-paying-the-price-for-benefits.asp
7. *Loc. cit.*
8. Webber, A. *Danger: Toxic Company*, Fast Company, Issue 19, November 1998, http://www.fastcompany.com/magazine/19/toxic.html, pp 152

9. Amble, B. *Employee disengagement a global epidemic*, Management Issues News, 16 November 2005, http://www.management-issues.com/display_page.asp?section=research&id=2772
10. Wallis, C. *The Multitasking Generation*, Time Magazine, 27 March 2006, pp 48-55, http://www.time.com/time/magazine/article/0,9171,1174696,00.html
11. McWilliams, G. *The Laptop Backlash*, The Wall Street Journal, 14 October 2005, pp B1, B5
12. Wallis, *Loc. cit.*
13. Trunk, P. *The ladder isn't the only way up*, 19 February 2006, http://blog.penelopetrunk.com/2006/02/19/the-ladder-isnt-the-only-way-up/

Chapter 4

1. National Center on Education and the Economy, *Executive Summary: Tough choices or tough times*, The Report of the New Commission on the Skills of the American Workforce, 2006, http://www.skillscommission.org/pdf/exec_sum/ToughChoices_EXECSUM.pdf, pp 8
2. National Academy of Engineering, *The Engineer of 2020: Visions of Engineering in the New Century*, 2004, http://books.nap.edu/catalog.php?record_id=10999, pp 54-56
3. Costlow, T. *Engineers take a hard look at soft skills*, EE Times India, 1 December 2000, http://www.eetindia.co.in/ART_8800388113_1800007_TA_60d17fe8.HTM
4. Gallo, C. *Lose the jargon or lose the audience*, Business week online, 1 December 2005, http://www.businessweek.com/smallbiz/content/nov2005/sb20051130_272052.htm
5. Bellis, M. *The history of the telephone*, accessed 16 Sep 2006, http://inventors.about.com/library/inventors/bltelephone.htm
6. Ament, P. *Mobile telephone*, accessed 16 September 2006, http://www.ideafinder.com/history/inventions/mobilephone.htm
7. **President's Commission on the Implementation of United States Space Exploration Policy**, June 2004, ISBN 0-16-073075-9, http://www.nasa.gov/pdf/60736main_M2M_report_small.pdf#search='aldridge%20commission%20document
8. *ibid.*, pp 50
9. *ibid.*, pp 7
10. The International Council on Systems Engineering (INCOSE) and the American Institute of Aeronautics and Astronautics System Engineering Technical Committee (AIAA SETC), *Systems Engineering*, August 1997, http://www.incose.org/ProductsPubs/pdf/SEPrimerAIAA-INCOSE_1997-08.pdf#search='why%20systems%20engineering
11. Symon, K. *Lack of skilled graduates hitting Scottish growth*, Sunday Herald, http://findarticles.com/p/articles/mi_qn4156/is_20041128/ai_n12592384/?tag=content;col1, 28 November 2004
12. Louk, D. *The new trend: Death by multitasking*, The Stanford Daily, 20 April 2006
13. Strayer, D. et al *Does cell phone conversation impair driving performance?*, http://www2.nsc.org/issues/idrive/inincell.htm, 13 March 2002
14. Atkinson, J. *Better Time Management*, Thorsons Publishing, 1992, ISBN 0-7225-2611-3, pp 20

Chapter 5

1. Selinger, C. *Learning where the jobs are*, IEEE Spectrum, http://spectrum.ieee.org/at-work/tech-careers/learning-where-the-jobs-are/0, 8 February 2006

2. Alton, J. *San Francisco shows how to lure biotech*, The Arizona Republic, 16 June 2005, http://www.azcentral.com/business/columns/articles/0616talton16.html
3. Phillips, R. *UK technology transfer training*, http://sciencecareers.sciencemag.org/career_development/previous_issues/articles/2006_02_03/uk_technology_transfer_training/(parent)/158, 3 February 2006
4. Associated Press (AP), *Fertilizer business sprouts from worm waste*, http://www.sptimes.com/2006/01/03/Business/Fertilizer_business_s.shtml, 3 January 2006
5. Demos, T. *Chevron CEO: Corn is not the answer*, http://money.cnn.com/2006/06/23/news/companies/chevron.fortune/index.htm, 23 June 2006
6. *Brazil leads the world in Ethanol Production*, accessed 8 August 2006, www.newenergyreport.org
7. Martinot, E. *Renewables 2005: Global Status Report*, Worldwatch Institute, http://www.ren21.net/pdf/RE2005_Global_Status_Report.pdf, 6 November 2005, pp 6
8. Associated Press (AP), *Nissan CEO: Hybrids showing sales slowdown*, http://msnbc.msn.com/id/12285537/, 12 April 2006
9. www.onstar.com
10. Stern, A. *One in four US bridges need repair – report*, http://www.reuters.com/article/idUSN2820066120080728, 28 July 2008
11. Associated Press (AP), *Doubt, Pain Linger 1 year after Minn. Bridge Fall*, http://cbs5.com/national/bridge.collapse.anniversary.2.785531.html, 1 August 2008
12. American Association of State Highway and Transportation Officials (AASHTO), *Bridging the Gap*, http://www.transportation1.org/BridgeReport/docs/BridgingtheGap.pdf, July 2008, pp 2
13. Stone, R. *Human Resource Management*, 5th Edition, Wiley & Sons Ltd, 2005, ISBN 0-470-80403-3, pp 4
14. May, T. *HR is out of sync with IT work*, http://www.computerworld.com/s/article/109645/HR_Is_Out_of_Sync_With_13_IT_Work, 20 March 2006
15. *Ibid. 13.* at pp 226

Chapter 6

1. Pencinger, C. *Loc. cit.*
2. The Personality Page, www.personalitypage.com, accessed 15 August 2006
3. Prof Ray Findlay's presentation to the IEEE 2005 leadership conference, St Louis, 2005
4. Telecon R. Lumsden/G. Walters, 27 March 2008
5. Telecon R. Lumsden/G. Walters, 27 March 2008
6. Withers, S. *Is there really a skills shortage?*, Voice & Data magazine, August 2006, pp 22-24
7. Patel-Predd, P. *What's up, Postdoc?* IEEE Spectrum, September 2006, http://spectrum.ieee.org/at-work/tech-careers/whats-up-postdoc, pp 60
8. *Ibid.*, pp 62
9. *Ibid.*
10. Perman, S. *Encouraging entrepreneurship at work*, Business Week Online, http://www.businessweek.com/smallbiz/content/aug2006/sb20060809_083684.htm, 9 August 2006
11. Marken, A. *Engineer to CEO*, Marken Communications, http://www.markencom.com/docs/01mar19.htm
12. Rothman, W. *Nike + iPod sport kit*, Time magazine, http://www.time.com/time/business/article/0,8599,1216589,00.html, 19 July 2006
13. Mettler, A. *Europeans must embrace entrepreneurship*, Business Week Online,

http://www.businessweek.com/globalbiz/content/jul2006/gb20060711_113037.htm, 11 July 2006

Chapter 7

1. Tufte, E. *Powerpoint is evil*, Wired magazine, http://www.wired.com/wired/archive/11.09/ppt2.html, Issue 11.09, September 2003
2. Marcus, R. *Powerpoint—Killer App?*, The Washington Post, 30 August 2005, http://www.washingtonpost.com/wp-dyn/content/article/2005/08/29/AR2005082901444.html, pp A17
3. Tufte, *Loc. cit.*
4. *The Provincial Letters*, http://en.wikipedia.org/wiki/Blaise_Pascal, accessed 22 April 2006
5. Taub, S. *Raytheon Chief Penalized for Plagiarism*, CFO Magazine, http://www.cfo.com/article.cfm/6878423/c_6876075?f=todayinfinance_next, 3 May 2006
6. Tung, B. *Our understanding of the solar systems took some unplanned detours*, 12 November 2001, http://www.strangehorizons.com/2001/20011112/kepler.shtml

Chapter 8

1. http://americanjobsfilm.com/GSBobEdwardspt2.mp3, accessed 1 October 2006
2. Jenkins, M. *Trans Siberian Express*, Boeing Frontiers Online, http://www.boeing.com/news/frontiers/archive/2003/october/i_atw.html, October 2003, Volume 2, Issue 6
3. Hearn, J. *Outsourcing is bad, insourcing is better*, 9 March 2004, www.thehill.com
4. ..., *Study says US tech hiring grows*, CNN Money.com, http://money.cnn.com/2006/02/23/news/economy/jobs_it_offshoring/index.htm, 23 February 2006
5. Friedman, T., *Learning to keep learning*, http://query.nytimes.com/gst/fullpage.html?res=9507E7DC1531F930A25751C1A9609C8B63, NY Times, 13 December 2006
6. Fishman, C. *No Satisfaction*, Fast Company, December 2006/January 2007, http://www.fastcompany.com/magazine/111/open_no-satisfaction.html, pp 84, 86
7. Committee for Economic Development of Australia (CEDA), Press Release, April 2004
8. Rothman, A. *Airbus vows software will speak same language*, http://seattlepi.nwsource.com/business/287165_airbussoftware02.html?source=rss, 2 October 2006
9. Soeiro, A. *The Bologna Process and the Globalisation of Engineering*, 7th WFEO Congress on Engineering Education Proceedings, pp 17
10. www.washingtonaccord.org
11. Boshier, J. *Creating International Mobility*, Engineers Australia, July 2003, pp 51
12. Wade, M. *Next step in globalisation: the workers*, Sydney Morning Herald, http://www.smh.com.au/news/national/next-step-in-globalisation-the-workers/2006/09/03/1157222010791.html, 4 September 2006
13. Smith, G. *IRCSET unveils research scheme worth €4.6m*, http://www.siliconrepublic.com/innovation/item/5541-ircset-unveils-research-sch, 5 January 2006
14. Skelly, B. *SFI to fund €11.7million software centre*, http://www.siliconrepublic.com/innovation/item/5332-sfi-to-fund-11-7m-br-soft, 9 November 2005
15. Kennedy, J. *Google to forge links with Irish universities*, http://www.siliconrepublic.com/business/item/5492-google-to-forge-links-with, 14 December 2005
16. Lucky, R. *The Impermanence of Knowledge*, IEEE Spectrum, March 2004, http://www.boblucky.com/reflect/mar04.htm, pp 56
17. Smith, C. S. *Going for your life*, APESMA Professional Network, April/May 2004, pp 14-15
18. Hardie, N. *Contemporary People Management 402*, La Trobe University MBA unit, pp 3.3

19. Conversation with General S. Worden (Rtd) (Ph.D), 2002
20. Topham, G. **Work to rule**, Sydney Morning Herald, http://www.smh.com.au/news/national/work-to-rule/2006/02/05/1139074110009.html, 6 February 2006
21. De Botton, A. http://www.alaindebotton.com/status.asp
22. De Botton, A. **Status Anxiety**, Channel 4, 2004
23. Karlin, S. **Young inventors of the world unite**, IEEE Spectrum, March 2004, http://spectrum.ieee.org/geek-life/profiles/young-inventors-of-the-world-unite, pp 47
24. Ibid.
25. Kumagai, J. **The Whistle-Blower's dilemma**, IEEE Spectrum, April 2004, http://spectrum.ieee.org/at-work/tech-careers/the-whistleblowers-dilemma, pp41-42
26. Tamara, P. **Sexing the tech**, The Age, http://www.theage.com.au/news/Management-Focus/Sexing-the-tech/2005/05/30/1117305545146.html, 27 May 2005
27. Search term 'sustainability', http://en.wikipedia.org/wiki/Main_Page
28. Ehrenfeld, J. R. **Feeding the beast**, Fast Company, Dec 2006/Jan 2007, http://www.fastcompany.com/magazine/111/next-essay.html, pp 42-43
29. http://www.newairplane.com/environment/ourenvironmentalcommitment/
30. Slackman, M. **The fine art of hiding what you mean to say**, The New York Times, http://www.nytimes.com/2006/08/06/weekinreview/06slackman.html?ei=5090&en=c34aa41fbe95aea9&ex=1312516800&pagewanted=all, 6 August 2006
31. Conversation with Lizzie Lee, August 2006
32. Agence France-Presse, **China to boost Science and Tech Spending to 20%**, http://www.spacewar.com/reports/China_To_Boost_Science_And_Tech_Spending_By_20_Percent.html, 13 March 2006
33. Kumagai, J., Hood, M. **China's tech revolution**, IEEE Spectrum, June 2004, http://spectrum.ieee.org/computing/networks/chinas-tech-revolution, pp 30

Chapter 9

1. http://www.cbs.com/primetime/numb3rs/about.shtml

Chapter 10

1. ..., Mayo Clinic HealthQuest, October 2006, ISSN 1092-1737, pp 8
2. Colvin, G. **What it takes to be great**, CNNMoney.com, http://money.cnn.com/magazines/fortune/fortune_archive/2006/10/30/8391794/index.htm, 17 Oct 2006
3. Friedman, T. **The example set by Armstrong ignored**, San Jose Daily News, 27 July 2005, pp 13
4. Poletti, T. **My fellow job seekers...**, San Jose Mercury News, 11 May 2005, pp 1, 16A
5. Roan, S. **To sharpen the brain, first hone the body**, Los Angeles Times, 9 January 2006, pp F4
6. Ibid.
7. Ibid.
8. Butler, C. **Running the Company**, Runners World, November 2005, pp 76
9. Ibid.
10. Levine, I. S. **Forty Winks: Science and Sleep**, http://sciencecareers.sciencemag.org/career_development/previous_issues/articles/2006_07_28/forty_winks_science_and_sleep/(parent)/68, 28 July 2006
11. IEEE Santa Clara Valley career fair, April 2005

Further Reading

Most books on the list below can be found through a quick search on Amazon.com.

Engineering Advice

1. Ashby, D. *Electrical Engineering 101: Everything you should have learned in school but probably didn't*, Newnes, 2008, ISBN 978-1856175067, 320 pges
2. Augustine, N. *Augustine's Laws*, AIAA, 6th edition, 1997, ISBN 978-1563472404, 395 pges
3. Baine, C. *Is There An Engineer Inside You?: A Comprehensive Guide To Career Decisions In Engineering*, 3rd Edition, Professional Publications, 2nd Edition, 2004, ISBN 978-1591260202, 182 pges
4. Baine, C. *High Tech Hot Shots—Careers in Sport Engineering*, NSPE, 2004, ISBN 978-0915409235, 144 pges
5. Berson, B. R. et al *Career Success in Engineering*, Kaplan AEC Education, 2007, ISBN 978-1419584398, 336 pges
6. Borchardt, J. *Career Management for Scientists and Engineers*, 1st Edition, American Chemical Society, 2000, ISBN 978-0841235250, 272 pges
7. Echaore-McDavid, S. *Career Opportunities in Engineering*, Checkmark Books, 1st Edition, ISBN 978-0816061532, 2006, 272 pges
8. Fouke, J. *Engineering Tomorrow : Today's Technology Experts Envision the Next Century*, Wiley-IEEE Press, 1st Edition, 2000, ISBN 978-0780353626, 324 pges
9. Galloway, P. *The 21st-Century Engineer: A Proposal for Engineering Education Reform*, American Society of Civil Engineers, 2007, ISBN 978-0784409367, 152 pges
10. Gardner, B. *Hired Minds: A Career Guide for Engineering Students and Graduates*, AIAA, 1st Edition, 2007, ISBN 978-1563478765, 137 pges
11. Garner, G. *Careers in Engineering*, McGraw-Hill, 3rd Edition, 2002, ISBN 978-0071545556, 192 pges
12. Garner, G. *Great Jobs for Engineering Majors*, Mc-Graw Hill, 3rd Edition, 2008, ISBN 978-0071493147, 192 pges
13. Goldberg, D. *Lifeskills and Leadership for Engineers*, McGraw Hill, 1997, ISBN 978-0074631546, 150 pges
14. Hutson, M. *Totally Amazing Careers in Engineering*, Sally Ride Science, 1st Edition, 2007, ISBN 978-1933798042, 32 pges
15. Kamm, L. *Real World Engineering—A guide to achieving career success*, Wiley-IEEE Press, 1991, ISBN 978-0879422790, 246 pges
16. Landis, R. *Studying Engineering: A Road Map to a Rewarding Career*, Discovery Press, 2nd Edition, 2000, ISBN 978-0964696952, 286 pges

17. Longuski, J. *Advice to Rocket Scientists: A Career survival guide for scientists and engineers*, AIAA, 2004, ISBN 978-1563476556, 84 pges
18. McCarthy, S. *Engineer your way to success*, NSPE, 2nd Edition, 2002, ISBN 978-0915409174, 101 pges
19. Roadstrom, W. H. *Excellence in Engineering*, Wiley, 1967
20. Selinger, C. *Stuff You Don't Learn in Engineering School: Skills for Success in the Real World*, Wiley-IEEE Press, 2004, ISBN 978-0471655763, 192 pges

Reports

1. National Academy of Sciences, Engineering and Institute of Medicine, Rising above the gathering storm—Energizing and Employing America for a Brighter Economic Future, 2007, ISBN 978-0-309-10039-7, www.nap.edu/catalog/11463.html, 590 pges
2. National Academy of Engineering, Educating the Engineer of 2020: Adapting Engineering Education to the New Century, 2005, ISBN 978-0-309-09649-2, www.nap.edu/catalog/11338.html, 208 pges
3. National Academy of Engineering, The Engineer of 2020: Visions of Engineering in the New Century, 2004, ISBN 978-0-309-09162-6, www.nap.edu/catalog/10999.html, 118 pges
4. National Center on Education and the Economy, Tough choices or tough times, 2006, ISBN 978-0787995980, http://www.skillscommission.org/?page_id=285, 208 pges
5. National Center on Education and the Economy, America in the Global economy, 2006, http://www.ncee.org/wp-content/uploads/2010/04/AmericaInGlobalEconomy.pdf, 65 pges
6. The Millennium Project; University of Michigan, Engineering for a changing world, 2008, http://milproj.ummu.umich.edu/publications/EngFlex_report/index.html, 131 pges
7. Deloitte Research, It's 2008: Do you know where your talent is?, 2004, http://www.dnlglobal.com/includes/repository/newsitem/Deloitte%20scenario%202008.pdf, 20 pges

Websites

1. American Association of Engineering Societies (http://www.aaes.org/)
2. American Institute for Aeronautics and Astronautics (www.aiaa.org)
3. American Society for Engineering Education (http://www.asee.org/)
4. American Society for Mechanical Engineers (www.asme.org)
5. Digest of Key Science and Engineering Indicators 2008 (http://www.nsf.gov/statistics/digest08/start.htm)
6. Engineering Workforce Commission (http://www.ewc-online.org/index.asp)
7. Grand Challenges for Engineering (http://www.engineeringchallenges.org/)

8. Institute of Electrical and Electronics Engineers (www.ieee.org)
9. Junior Engineering Technical Society (www.jets.org)
10. National Society of Professional Engineers (www.nspe.org)
11. New Commission on the Skills of the American Workforce (http://www.skillscommission.org)
12. Society of Automotive Engineers (www.sae.org)
13. World Federation of Engineering Organizations (http://www.wfeo.org/index.php)
14. US Department of Labor: Engineers (http://www.bls.gov/oco/ocos027.htm)

DVD

1. Spotts, G. *American Jobs*, Spottsfilm, 2005, 62 mins
2. De Botton, A. *Status Anxiety*, Channel Four Television Corporation, 2004, 148 mins
3. Morrow, J. *Declining by degrees—Higher Education at Risk*, www.decliningbydegrees.org, 120 mins

Bibliography

1. Claxton, G. ***Wise-Up—The Challenge of Lifelong Learning***, Bloomsbury, 1999, ISBN 1-58234-092-7, 374 pges
2. Collin, A., Young, R. ***The Future of Career***, Cambridge University Press, 2000, ISBN 0 521 64965 X, 321 pges
3. Fiorina, C. ***Tough Choices***, Portfolio, 2006, ISBN 1-59184-133-X, 319 pges
4. Florida, R. ***The Flight of the Creative Class***, Harper Business, 2005, ISBN 0-06-075690-X, 326 pges
5. Friedman. T. L. ***The World is Flat***, Farrar, Straus and Giroux, 2005, ISBN 978-0-374-29288-1, 488 pges
6. Fox, C., Trinca, H. ***Better than Sex—How a whole generation got hooked on work***, Random House Australia, 2004, ISBN 1 74051 196 4, 226 pges

About the Author

Reece Lumsden has 20 years professional experience in the Aerospace and Defense sectors and has experienced the working environments of Korea, Australia, Mexico, France and India. A two time applicant for NASAs Astronaut candidate program, Reece is currently a Project Engineer for a large Aerospace company in Everett, Washington.

Reece holds a Bachelors of Electrical and Electronic Engineering from The University of Western Australia (UWA), a Masters of Space Studies from the International Space University (ISU) in France, a Masters of Business Administration with a focus on Technology Management from The Chifley Business School, La Trobe University, a Masters of Science in Systems Engineering with a major in Network Centric Systems from the Missouri University of Science and Technology (MS&T) and is currently undertaking a Doctorate of Engineering in Engineering Management and Systems Engineering with Old Dominion University (ODU). In addition, he holds Master certificates in Applied Project Management and Six Sigma from Villanova University, a Master Certificate in Supply Chain Management from the University of San Francisco (USF), and a post-graduate certificate in System Dynamics from the Universite Politechnica de Catalunya (UPC), Spain. He also holds a Project Management Professional (PMP) certification with the Project Management Institute (PMI), has completed the Advanced Project Management Program with Stanford Universities Center for Professional Development and has completed the Massachussets Institute of Technology (MIT) Architecture and Systems Engineering Certificate Program.

Reece was awarded the Korean Association of Science and Technology in Australia (KASTA) award in 1994, during which time he interned with Samsung Data Systems (SDS) in Seoul, Korea, the Young Professional Engineer of the Year award in 2004 from the Australian Capital Territory Division of Engineers Australia, and in 2008, was one of 92 young engineers under the age of 40, selected from across the United States by the National Academy of Engineering (NAE) to attend the U.S. Frontiers of Engineering symposium.

Reece's focus on Project Management and Systems Engineering has been developed through numerous career episodes; as an officer in the Royal Australian Air Force (RAAF) involved in unmanned aerial vehicles trials, future space concepts policy development and enterprise architectures for capability development, as a consultant and Senior consultant with a private Defense services firm to the Australian Defense Force (ADF) working various capability programs, as a consultant to the National Aerospace Development Center (NADC), executing workforce development initiatives as part of a US Department of Labor grant and finally as an engineer working Systems Integration issues across a variety of airplane programs (787 Dreamliner, 777X and KC-46 'Pegasus' Tanker) and strategy concerns across various business units (Electrical Design and Supply Chain Strategy).

Reece has served on the executive committee of the Australian Space Industry Chamber of Commerce (ASICC), as the Chapter Chair for the Aerospace and Electronic Systems Society (AESS) and as a member of the Peacekeepers Taskforce, Space Policy Summit, held at the Baker Institute, Rice University in 2001. He was elected to the AESS Board of Governors in 2010 and served as their representative to IEEE-USAs Research and Development Policy Committee. He has also served on the American Institute for Aeronautics and Astronautics (AIAA) Systems Engineering Technical Committee since 2006.

Reece is a lifetime member of AIAA, a Senior Member of the IEEE and is a member of Engineers Australia, the Association of Professional Engineers Scientists and Managers Australia (APESMA) and the Project Management Institute (PMI).

Reece has contributed articles on engineering, the space industry and career guidance for over a decade to such publications as *IEEE Spectrum, Aerospace America, Engineers Australia* magazine, *Space Times, APESMA Professional Network* magazine and *Today's Engineer*. *The View From Here* is Reece's second book, and first as a sole author. His first book was as a contributing author to *The Athena Global Earth Observation Guide 2005*, ISBN 0-9737106-0-8

Outside of work, Reece is an avid runner, having completed almost 120 events including the 2011 Boston Marathon and as part of a 6 person 'Ultra' team to complete a 200+ mile relay. He also holds a 2nd Dan Black Belt in WTF style Taekwondo. Reece lives with his wife Nicole in Shoreline, Washington where they enjoy their 9 year old Australian Labradoodle named Humphrey.

www.ingramcontent.com/pod-product-compliance
Lightning Source LLC
Chambersburg PA
CBHW020652220526
45464CB00001B/407